MINES AND MINERAL RESOURCES O   I0052432   COUNTY, CALIFORNIA

By Denton W. Carlson * and William B. Clark *

OUTLINE OF REPORT

Illustrations

* Junior Mining Geologist, California Division of Mines. Manuscript submitted for publication January 1953.

ABSTRACT

Amador County, 598 square miles in extent, lies largely in the Sierra Nevada. The crest of the Sierra Nevada, which, in Amador County, attains an altitude of 9,371 feet, is in the eastern part, while the western part borders the alluvial plain of the San Joaquin Valley.

The bedrock of the Sierra Nevada consists of steeply dipping, isoclinally folded and faulted metamorphic rocks which have been invaded by several types of igneous rocks. Overlying the bedrock are nearly flat-lying Tertiary sediments and volcanic rocks. The pre-Tertiary rocks are composed of Carboniferous, Permian, and Jurassic metasediments and volcanics and late Jurassic igneous rocks, chiefly granodiorite. The Tertiary rocks are Eocene sands, clays, and auriferous gravels and Miocene and Pliocene volcanic rocks.

The chief industries of Amador County are logging, agriculture, and mining. Since 1880 more than $165,500,000 worth of mineral products has been produced. The single commodity responsible for nearly 85 percent of this total is gold, most of which has been produced from that part of the Mother Lode gold belt between Jackson and Plymouth. Actually, this segment has been the richest part of the entire 115-mile length of the lode. Other important mineral products are clay, coal (lignite), and copper.

The largest producers of gold have been the Kennedy, Argonaut, Keystone, and Plymouth mines, all of which have been inactive for nearly a decade. The Central Eureka mine which is now active is one of the few Mother Lode gold mines which survived the government restriction order L-208 of 1942 and increased mining costs. Other past important gold producers have been the Fremont-Gover, Lincoln, Zeila, Oneida, Original Amador, South Eureka, Wildman, Mahoney, Treasure, and Bunker Hill mines. In the east belt the Belden, Rainbow, and Pioneer Lucky Strike mines were important producers. Several small east belt mines are now active. Substantial quantities of gold have been produced by gold dredges, hydraulic mines, drift mines, and by the re-working of old tailings. Several small dragline dredges are now active near Volcano.

Clay is produced chiefly from Eocene beds in the western part of the county. The chief clay producers are Gladding McBean and Company, Western Refractories Company, and the Pacific Clay Products Company. Coal is now produced in the Buena Vista area by the American Lignite Products Company and the Humacid Company. Two mines, the Newton and Copper Hill, have produced most of the copper in Amador County.

Other mineral products of Amador County are silver, chrome, manganese, iron ore, asbestos, sand and gravel, crushed rock, limestone and marble.

INTRODUCTION

Amador County is an irregular-shaped area lying between the Mokelumne River on the south and the Cosumnes River on the north. It extends from Alpine County on the east to Sacramento and San Joaquin Counties on the west. Originally part of Calaveras County, Amador County was created on May 11, 1854. It was named for Jose Maria Amador, son of Sergeant Pedro Amador, a Spanish soldier who settled in California in 1771.

The first authentic report of the presence of white men in Amador County was in 1846, when Captain John Sutter, accompanied by a small party of Indians and a few white men, conducted logging operations on the ridge between Sutter and Amador Creeks. A few months after the discovery of gold 45 miles to the north in 1848 prospectors began to swarm into what is now Amador County. By 1851 towns such as Drytown, Fiddletown, Jackson, and Volcano had been established.

Geography

Amador County embraces an area of 598 square miles. Altitude varies from 200 feet in the western part of the county to 9,371 feet, the altitude of Mokelumne Peak, in the east. The central area averages about 4,000 feet.

In 1950 the population of Amador County was 9,151. Jackson, the county seat, is the largest town, with a population of 1,879. Other important towns are: Ione, 1,071, Sutter Creek, 1,151, Plymouth, 382, and Amador City, 151. The county is served by the Ione branch of the Southern Pacific railroad and the Amador Central railroad which connects with the Southern Pacific at Ione and extends to Martell. State Highway 49 runs north-south across the central portion of the western end. State Highway 88 extends the length of the county in a general east-west direction.

In the western foothills the climate is hot during the summer and mild during the winter. The central portion has a moderate climate but the most mountainous eastern portion has heavy snows in the winter. Rainfall averages 30 inches in the western portion and 43 inches in the eastern portion.

Amador County is drained by the Mokelumne River in the southern part of the county and the Cosumnes River along the northern margin. Other important streams are Jackson, Sutter, Rancheria, and Dry Creeks. Although the drainage pattern is partly controlled by the north to northwesterly trend of the major geologic structures, the main rivers and streams flow in a general southwesterly direction.

Industries

Logging, agriculture, and mining are the principal industries of Amador County. Logging operators cut trees in 1950 worth $9,000,000. The value of agricultural products in 1950, primarily livestock, amounted to $3,100,000. Mining operations, the third most important industry of Amador County, produced mineral products worth $1,026,560 in 1950. Hydroelectric power plants on the Mokelumne River are operated by the Pacific Gas and Electric Company.

GEOLOGY

Amador County lies almost entirely in the Sierra Nevada geomorphic province; only the extreme western portion lies in the Great Valley. From the Great Valley eastward, the range gradually rises to the gla-

FIGURE 1.   Logtown Ridge formation, showing jointing in porphyritic andesite exposed in road cut 4 miles east of Ione.

ciated crest in the vicinity of Mokelumne and Thimble Peaks, both of which lie above 9000 feet.

The older rocks of the Sierra Nevada, commonly called the "Bedrock series," consist of isoclinally folded, complexly faulted metamorphic rocks of Paleozoic and Mesozoic ages, intruded by several types of igneous rocks, chiefly granitic. Unconformably overlying these rocks in the western portion of Amador County are much younger, nearly flat-lying Tertiary sediments. These nearly flat-lying sediments are commonly called "Superjacent series."

In Amador County, the older metamorphic rocks are divided into the Calaveras and Amador (Taliaferro, 1943, pp. 282-284) groups and Mariposa formation. The Calaveras group includes all of the pre-Mesozoic rocks in this county while the Amador group and Mariposa formation are Jurassic. Taliaferro has divided the Amador group of Amador County into two distinctive formations: the Cosumnes and Logtown Ridge.

Rock Formations

Calaveras Group (Carboniferous and Permian in Part). The Calaveras group is an undifferentiated suite of rocks ranging mainly from Carboniferous to Permian; part of the series may be older than the Carboniferous. In this region, it consists predominantly of highly contorted, poorly bedded, blackish-gray, recrystallized cherts and quartzites. Present in smaller amounts are greenstones, low-grade carbonaceous slates, mica schist, and grayish-colored recrystallized limestones which in rare cases contain crinoid and coral debris. These rocks most commonly strike north-northwest with a high angle of dip to the east.

Amador Group (Middle to Upper Jurassic). The lowest formation of the Amador group, the Cosumnes, lies unconformably on the Calaveras.[1] Sheared sands, grits, slates, and a thick basal conglomerate containing pebbles and boulders of chert, quartz, schist, and volcanic rocks in a sandy or gritty matrix characterize the Cosumnes. Some of the sheared sands have been completely metamorphosed to mica and sericite schists.

The Logtown Ridge formation is composed of massive porphyritic augite andesite with a coarse andesite agglomerate at the base. The andesite commonly has a fine-grained groundmass and contains coarse phenocrysts of brownish-black augite. The agglomerate occurs as thick beds and contains angular fragments of andesite, quartz, chert, and schist. Interbedded in smaller amounts are welded tuff, tuffaceous sandstone, and slate.

Mariposa Formation (Upper Jurassic). A conspicuous formation, the Mariposa consists of extensive beds of dark slates with small amounts of phyllite and schist and a thick, coarse basal conglomerate. The slates are bluish- to black-gray in color and crop out extensively as nearly vertical beds in Amador County. Small cross-bedded sand lenses are common in these slates. The coarse basal conglomerate contains pebbles and cobbles of chert, quartz, volcanic rocks, and schist.

Ione Formation (Middle Eocene). The Ione, which is middle Eocene in age, occurs on the low rolling hills of western Amador County in nearly flat-lying beds. It consists essentially of soft, friable white sand with scattered grains of mica and interbedded lenses of white clays and lignite. Also occuring in the Ione are the gold-bearing Eocene stream gravels composed of pebbles and boulders of quartz, chert, granites, and volcanic rocks.

Valley Springs Formation (Middle (?) Miocene). The Valley Springs formation as mapped by Piper, Gale, et al., (1939, pp. 71-80) is composed predominantly of fragmental and rhyolitic detritus of Miocene age. This formation is composed of rhyolite tuff, breccia, conglomerate, and siltstone with a sandstone member at the base. It occurs west of and in the vicinity of Buena Vista Peak in the southwestern part of the county.

Mehrten Formation (Upper (?) Miocene and Pliocene (?)). The Mehrten formation, which unconformably overlies the Valley Springs formation, consists largely of volcanic debris laid down by the Mio-

---

[1] Taliaferro, N. L., personal communication, 1949.

FIGURE 2. Roadcut exposing Mehrten formation on Jackson-Ione road.

Pliocene inter-volcanic river systems. This formation, which was named and mapped by Piper, Gale, et al., (1939, pp. 61-71,) is composed chiefly of boulders and fragments of porphyritic andesite with smaller amounts of siltstone, sandstone, clay, and conglomerate.

Recent Alluvium. Recent alluvium consists of sand, silt, and gravel in and adjacent to the present stream channels.

Intrusive Rocks

During late Jurassic time the crust beneath Amador County was intruded by granitic rocks of the Sierra Nevada batholith. This batholith, which is predominantly granodiorite, also included other rocks, such as granite, quartz monzonite, gabbro, feldspar and quartz porphyries, diorite, and pegmatite. Much of the basic rock has since been altered to serpentine. The weathered products from these intrusive igneous rocks constituted a major part of detritus that built the later Tertiary formations.

Geologic History

The oldest rocks found in Amador County are those of the Calaveras group which were deposited as marine sediments late in the Paleozoic era approximately 230 to 255 millions of years ago. These sediments were deposited as mud, sand, and marl which have since been changed into the different metamorphic rock types of the Calaveras group. Also associated with these rocks and with the submarine volcanic products of that time are chemically deposited manganese-bearing, siliceous sediments. During the Paleozoic a vast open sea covered what is now Amador County. At or near the end of the Paleozoic a crustal disturbance took place which resulted in at least partial destruction of the sea basin and creation of a mountain chain of unknown extent.

Between the end of the Paleozoic era and the middle of the Jurassic period the sea again advanced over what is now Amador County. Near the end of the middle Jurassic period, sediments (Cosumnes member of the Amador group) were deposited at the bottom of this Mesozoic sea to the extent of many thousands of feet. Marine sedimentation was followed by submarine volcanism, the products of which contributed to the Logtown Ridge member of the Amador group. Subsequent marine deposition gave rise to what is now termed the Mariposa formation. Associated with the accumulation of the Jurassic submarine volcanics are manganese deposits of limited size. These were laid down on the sea floor as chemically deposited sediment derived by precipitation from volcanic hot-spring water.

In late Jurassic time, orogenic processes began to elevate a new Sierra Nevada. This effected an almost complete withdrawal of the ancient sea from what is now the Sierra Nevada area. The nearly flat-lying Mesozoic sedimentary beds were folded into a complicated series of folds trending for the most part northwest, while the sediments themselves were being changed into metamorphic rocks. The ancestral Sierra Nevada of late Jurassic time was thus created as a fold-mountain range. Prior to destruction of the Jurassic sea-basin, igneous rocks of peridotite composition were intruded into the marine sediments as sills and irregular massive intrusions. Introduced with these peridotite rocks, now altered to serpentines, were segregations and disseminated patches of chromite.

Toward the end of the period of folding which elevated the ancestral Sierra Nevada, granitic rocks were introduced into the folded crust on a major scale, followed by hydrothermal deposition deep within the folded crust of gold, copper, and zinc. A very long period of weathering and erosion followed the elevation of the Jurassic Sierra Nevada with its attendant emplacement of granitic rocks and gold quartz veins. For twenty or thirty million years erosion stripped away the rocks lying above the gold deposits and cut deeply into the gold-bearing veins themselves. The elevations of the Jurassic Sierra Nevada were greatly reduced and broad river valleys developed. Thus the stage was set for the important and complicated series of events of the Tertiary which governed the character and location of the placer gold deposits destined to play such an important part in the human history of California.

Just before the opening of the Tertiary period, the Upper Cretaceous sea advanced east across a narrow belt of the ancestral Sierra Nevada marginal to the present Great Valley. Thin, relatively flat-lying fossiliferous marine deposits were laid down on the folded, eroded bedrock surface and patches of these old deposits are still found from place to place along the western edge of the present Sierra Nevada. In Amador County, however, Cretaceous deposits so far have not been found and were either removed by erosion or else were never deposited.

By Eocene time, the opening epoch of the Tertiary, the ancestral Sierra Nevada had been greatly worn down and the climate had changed from the temperate one of the Jurassic to subtropical. With the reduced elevations and the warm humid climate, chemical decay of the bedrock proceeded slowly but deeply. The finer products of this decay, chiefly clay and quartz sand, were removed by running water and deposited along the margins of the Eocene sea, giving rise to valuable clay and quartz sand deposits. This sea lapped onto the flanks of the

SUMMARY OF ECONOMIC GEOLOGY OF AMADOR COUNTY

Geologic age		Formations	Distribution	Rock types	Mineral deposits
QUATER-NARY		Alluvium	Stream channels, terraces	Sand, silt, gravel	Placer gold, sand and gravel
TERTIARY	Pliocene	Mehrten	Patches on hills in western and central portion	Flow andesite, andesitic conglomerate, silt, sand, clay	Potential source of stone
	Miocene	Valley Springs	Patches in southwest portion	Rhyolite tuff, breccia	Potential source of building stone
	Eocene	Ione	Extensive beds in western portion of Amador County	Clay, sand, gravel	Clay, lignite, sand, gravel, placer gold, dimension stone, impure iron ore
MESOZOIC		Veins	Mother Lode, East Belt, and Foothill Belt	Vein quartz	Gold, silver, copper, lead, zinc
	Jurassic	Granitic rocks	Extensive bodies cropping out in eastern and central portions of Amador County	Granodiorite, other intrusive rocks	Potential source of building stone and aggregate
		Serpentine	¼-to-½ mile-wide discontinuous belt trending NNW, between Ione and Highway 49	Serpentine altered from ultrabasic igneous rocks	Chromite, ornamental stone
		Mariposa	Two NW-trending mile-wide belts; one along Mother Lode, the other 5 to 6 miles west of Mother Lode	Slate, metaconglomerate, and residual clay	Building slate, clay
	Amador group	Logtown Ridge	NW-trending belts west of Mother Lode 100-500 feet wide	Meta-andesite, agglomerate tuff, and residual clay (lateritic)	Building stone, aggregate, clay (lateritic)
		Cosumnes	NW-trending belts west of Mother lode 100-500 feet wide	Metamorphosed sand, silt, slate, conglomerate	
PALE-OZOIC	Permian Carboniferous	Calaveras group	Extensive NW-trending belts west and east of Mother Lode	Metachert, quartzite, greenstone, limestone, slate	Marble, limestone, manganese; potential source of other varieties of building stone

FIGURE 3.   Glaciated granitic surface at east shore of Silver Lake.

ancestral Sierra Nevada in much the same fashion as had the Upper Cretaceous sea before it. An interfingering of Eocene river and marginal marine deposits resulted. The coarser, heavier materials, notably gold and quartz pebbles from the veins, resistant to decay and to abrasion, were left behind and became concentrated in the stream channels as placer deposits. These Eocene placer deposits were unusually rich and were characterized by an abundance of white quartz pebbles and a paucity of fresh volcanic debris. They were confined to channels worn into bedrock and in most places were overlain by white rhyolite ash, or pipe clay as it was called.

At or near the close of the Eocene epoch, volcanism which broke out in the Sierra Nevada was destined to have a profound effect upon the economic geology of the placer gold deposits. It began with falls of white rhyolite ash low down in the Sierra Nevada and with both flows and ash falls higher in the range. So much volcanic ash fell that it choked the stream channels and completely changed the drainage system of the Sierra Nevada. River gravels and placer deposits developing along water courses cut after the initial volcanism tended to be much leaner in gold and were characterized by an abundance of rhyolite pebbles. By late Miocene or early Pliocene time volcanic activity reached its peak with emission of massive flows of gray andesite as well as large volumes of fragmental material, the latter frequently giving rise to extensive mud flows. These volcanic products deeply buried the Eocene channel deposits and Miocene intervolcanic channel deposits alike, in many places to depths of several hundred feet. Channel deposits developing during or after the emission of andesite were still leaner in gold and commonly were not of economic value at all. Burial of Eocene gold placers by volcanic debris and by the later extruded latite and basalt flows did, however, serve to protect in some measure the placers from obliteration by later erosion.

The abundant semi-tropical flora which flourished along the margin of the Sierra Nevada contributed sufficient vegetal material to the swamp and marginal marine deposits to give rise to valuable beds of lignite now being exploited for natural waxes. These lignite deposits tend to be associated with other valuable deposits of clay and white quartz sand.

In the late Pliocene and early Pleistocene, major re-elevation of the Sierras by faulting took place, probably accompanied by violent earthquakes. Snow and ice packs on the peaks began to build up and migrate down the slopes as glaciers, elaborately carving the highland topography. Pleistocene glaciers occupied the northeastern tip of Amador County to a little south of Silver Lake, and just crossed into the southeast corner of the county.

Continued stripping of gold lodes in Pleistocene and Recent times enriched the present stream channels with placer gold. The glaciers retreated from Amador County and the Sierras in the early Recent epoch, and the present climate and topography developed.

METALLIC MINERALS

Chromite

Chromite deposits occur as magmatic segregations in ultrabasic igneous rocks or in serpentine derived from ultrabasic rocks. A north-northwest trending belt of serpentine ranging from $\frac{1}{4}$ to $\frac{1}{2}$ mile in width crops out discontinuously in western Amador County, particularly in the Ione area.

Most of the chromite deposits of Amador County occur as pods or irregular lenses associated with serpentine or serpentinized dunite. Some of the ore occurs as clots or bunches of disseminated low-grade chromite.

Minor amounts of chromite were produced in Amador County during World War I and some was produced in 1909. Since World War I, the mines of this county have been inactive. For more detailed information on chromite in Amador County, consult California Division of Mines' Bulletin 134 (Cater, 1948).

Copper

A recorded total of nearly eight million pounds of copper has been produced in Amador County since 1880. Prior to 1880, substantial amounts of copper were produced in this county during the state's first copper boom in the Civil War days of the early 1860's.

Irregular and almost continuous production of copper in Amador County began in 1895 and continued until the end of World War I. Most of the small amount of copper recovered between World Wars I and II was a by-product of gold mining; but World War II stimulated the county's industry. More than two million pounds of copper was produced during 1946, the peak year. Since 1947, only minor amounts of copper have been recovered.

Two mines are responsible for the largest portion of the total production of copper, the Copper Hill mine and the Newton mine. Bulletin 144, "Copper in California" (Jenkins, 1948a, pp. 214-216), contains a list of copper properties in Amador County.

During 1942-45 the U. S. Geological Survey and the U. S. Bureau of Mines, as part of the strategic minerals program, mapped and diamond-drilled some of the copper properties in this county.

A belt of rocks containing copper- and zinc-bearing minerals trends north to northwestward across the width of Amador County in the foot-hills west of the Mother Lode. Within this belt are lenticular sulfide bodies formed by replacement of wall rocks along the zones of shearing, crushing, and faulting in the Amador and Calaveras groups of rocks. Ore bodies are composed predominantly of pyrite and chalcopyrite with smaller amounts of sphalerite, bornite, chalcocite, tetrahedrite, galena, pyrrhotite, gold, and silver. In a few of these deposits sphalerite is a major ore mineral. Ore deposition was controlled primarily by structure and like the gold mineralization was genetically related to the emplacement of the granitic batholith.

Further details concerning Amador County copper may be found in California Division of Mines Bulletin 144 (Heyl, 1948, pp. 15-19).

Copper Hill Mine. Location: SE $\frac{1}{4}$ sec. 34, T. 8. N., R. 9 E., M.D.M., about $2\frac{1}{2}$ miles northwest of Four Corners. Ownership: W. F. Detert Estate, 1715 Mills Tower, San Francisco, California.

The first mining operations at the Copper Hill mine began in 1860-61 and the mine was in operation for over twenty years (Aubury, 1905, p. 186). Large amounts of copper ore and matte were shipped to be refined in Europe. The mine was re-opened by the late W. F. Detert shortly after 1902 and was shut down in 1911, according to Mr. Victor Bonnefoy.

Some work was done during 1943 in the Jackass shaft by J. P. Donovan of San Francisco but there is no recorded production from the mine for that year.

In April, 1948, this mine was again reopened by Victor Bonnefoy of Pine Grove and Clarence Miller of Petaluma who jointly leased the property. The mine is now leased by Miller and sub-leased to Ralph E. Fitzgerald and associates of Jamestown.

The ore consists of chalcopyrite and sphalerite with smaller amounts of chalcocite, bornite, pyrite, malachite, and azurite in a sheared and brecciated zone in Logtown Ridge metavolcanic rocks which strikes northwest and dips 60° northeast. A large cross-faulted zone, also heavily mineralized, cuts the vein at an acute angle, has a northwest strike and dips about 60° to the northeast. In 1943, the U. S. Bureau of Mines diamond-drilled this property. One drill hole northeast of the Jackass shaft encountered an ore body containing 22.37 percent zinc, 0.43 percent copper, 9.31 ounces of silver, and 0.02 ounce of gold.[2]

Three main inclined shafts which dip 60° to 70° northeast are on the property in addition to many smaller shafts and prospect holes. The mine was worked rather extensively during the early 1900's through the Main (700-foot) shaft. About 450 feet southeast of the Main shaft is the Hobo (450-foot) shaft and 900 feet north of the Main shaft is the Jackass (125-foot) shaft. Six levels off the Main shaft include extensive drifting, crosscutting, and stoping. The mine has about 5200 feet of underground workings.

At the present time, two men, including Roy Brown, mine superintendent, are surveying and reopening the Jackass shaft.

[2] West belt copper-zinc mines, Amador and Calaveras Counties, California: U. S. Bureau of Mines War Minerals Rept. 103, p. 8, 1943.

MINERAL PRODUCTION OF AMADOR COUNTY, 1880-1950

Year	Gold, value	Silver, value	Coal Tons	Coal Value	Copper Pounds	Copper Value	Pottery clay Tons	Pottery clay Value	Lime Barrels	Lime Value	Marble Cu. ft.	Marble Value	Brick M	Brick Value	Misc. Amount	Misc. Value	Substance
1880	$1,495,053	$1,953															
1881	1,450,000	1,500															
1882	1,500,000																
1883	1,590,000																
1884	2,000,000	2,000															
1885	2,145,591	3,700															
1886	1,874,062	6,136															
1887	1,979,956	2,049															
1888	1,750,000	3,500	24,404	$38,606													
1889	1,560,075	6,398	30,000	45,000													
1890	1,459,982	9,357															
1891	1,395,962	13,895	21,323	31,984													
1892	1,210,383	8,008															
1893	1,505,973	5,230															
1894	1,331,916	280	15,280	23,020			2,500	$3,000									
1895	1,391,929	1,099	21,323	31,985	16,500	$1,650	9,960	10,285			25,941	$35,826					
1896	1,523,351	3,767	19,776	29,662	30,000	3,000	8,413	27,825			4,864	6,566					
1897	1,324,472	3,477	20,000	25,000			3,492	9,540			4,389	5,415					
1898	1,806,363	1,742	18,500	29,550	3,000	300	7,197	8,297			3,864	6,280					
1899	1,844,868	6,902	18,500	23,125			10,700	10,900			2,850	3,594					
1900	1,373,788	14,915	27,477	41,215	220,000	34,100	11,500	9,100			4,582	7,925	600	$7,000			
1901	1,823,827	7,444	25,000	30,000	52,000	8,190	10,050	7,100			4,103	5,891					
1902	1,629,151	2,686	5,450	10,912	130,000	14,620	12,723	13,728									
1903	1,609,744	4,336			10,000	900	22,000	19,460			2,945	4,630					
1904	2,060,574	4,055			14,000	1,400	20,608	10,770			6,300	8,016					
1905	2,445,815	17,930			10,000	1,560	21,775	20,000	1,700	$1,700	3,074	5,379					
1906	2,260,373	14,579			8,648	1,669	26,789	28,119	1,000	1,500	4,785	6,558				750	Glass sand.
1907	2,116,182	13,515			5,300	1,020	12,465	13,992	1,000	1,200	2,703	3,950					
1908	1,876,175	13,239			53,940	3,440	23,322	25,369	800	960			2109	61,369	1,000 tons	1,200	Limestone.
															10 tons	1,600	Asbestos.
															1,072 lbs.	40	Lead.
1909	2,298,785	16,701			288,472	36,641	33,563	32,724	1,200	1,440			1429	28,572	1,000 tons	1,375	Limestone.
															2 tons	200	Asbestos.
															41 tons	332	Chromite.
1910	2,646,246	20,916			151,484	14,386	39,446	49,339	1,400	1,680			2000	30,000	1,000 tons	1,500	Limestone.
															10,100 tons	10,100	Quartz sand.
															11,200 cu.ft.	5,600	Sandstone.
1911	2,832,395	28,899			227,848	28,481	43,352	37,359	1,200	1,500			2000	20,000	600 tons	6,000	Soapstone.
															90,000 cu.ft.	45,000	Sandstone.
1912	2,796,194	32,037			175,608	28,975	35,100	36,856	800	1,040			2500	25,000	6,000 cu.ft.	3,000	Sandstone.
															700 tons	2,100	Soapstone.
																$318,422	Unapportioned, 1900–1909.
Totals								Miscellaneous stone, value			70,400	$100,030					

1913	2,901,898	18,097			19,023	2,949	39,678	38,653	1,000	1,200	670	2000	30,000	2,500 cu. ft.	2,500	Sandstone.
														350 tons	2,420	Soapstone.
														1,960 tons	3,556	Quartz.
														877 tons	670	Glass sand.
															11,237	Other minerals.
1914	3,082,002	17,032	5,700	10,062	5,251	694	32,223	33,114	1,540	2,008		2500	50,000	16,888 tons	9,855	Glass sand.
														44 lbs.	2	Lead.
														6,250 tons	2,400	Quartz.
														3,960 cu. ft.	1,500	Sandstone.
														610 tons	2,440	Soapstone.
1915	3,894,125	20,409			4,185	732	40,156	38,879	1,000	1,200	1,300	4000	80,000	523 lbs.	25	Lead.
														13,339 tons	16,142	Silica.
															10,950	Other minerals.
1916	3,660,550	18,705	[1]		12,349	3,038	29,246	31,106	[1]		1,300	[1] and tile		300 tons	3,700	Chromite.
														4,341 tons	12,802	Silica.
														495 tons	2,475	Soapstone.
															77,752	Brick, coal, lime, manganese, sandstone.
1917	3,664,164	21,358	[1]		19,352	5,283	28,970	28,625			1,200		95,345	65 tons	1,420	Chromite.
														4,771 tons	20,766	Silica.
															13,033	Coal, lead, manganese, platinum, soapstone, zinc.
1918	3,249,385	29,590	[1]		[1]		13,562	34,346			6,500		[1]	88 tons	4,400	Chromite.
														13,747 tons	61,724	Silica.
															66,695	Brick, coal, copper, manganese, mineral paint, platinum, soapstone.
1919	2,920,492	33,254	[1]				[1]						[1]		142,523	Clay and clay products.
														8,440 tons	67,366	Silica.
															9,953	Coal, manganese, platinum, sandstone, soapstone.
1920	1,788.793	19,780	[1]				25,719	61,808			680		[1]	6,116 tons	36,432	Silica.
															102,707	Brick, coal, mineral paint, platinum, soapstone.
1921	2,167,443	35,460					22,124	46,664			1,125		[1]	1,802 tons	20,646	Silica.
															97,126	Brick and platinum.
1922	2,241,100	32,287					39,572	68,126			7,300			865 tons	5,030	Silica.
															125,220	Other minerals.[3]
1923	1,734,133	15,153					45,887	58,196			28,515				119,877	Other minerals.[4]
1924	2,706,508	18,251	[1]		[1]		64,317	87,444			3,050				123,612	Other minerals.[5]
1925	2,338,101	16,123	[1]		[1]		63,889	95,946			31,100				11,003	Other minerals.[6]

MINERAL PRODUCTION OF AMADOR COUNTY, 1880-1950—Continued

Year	Gold, value	Silver, value	Coal Tons	Coal Value	Copper Pounds	Copper Value	Pottery clay Tons	Pottery clay Value	Lime Barrels	Lime Value	Miscellaneous stone Value	Brick M	Brick Value	Miscellaneous Amount	Miscellaneous Value	Substance
1926	2,167,275	13,422	1		1						24,900	1			237,792	Brick and clay (pottery).[6]
1927	1,922,714	11,319	1				118,636	165,210	1		10,400	1		1,267 lbs.	101 / 8,010 / 157	Lead. / Other minerals.[6] / Lead.
1928	2,236,922	14,317	1		1,402	202	96,209	116,000			189,900		1	2,494 lbs.	97,998 / 86,838	Other minerals.[7] / Brick, coal.
1929	1,601,881	9,392					60,487	88,846			696,500				101,618	Brick, coal, copper, lead, marble.
1930	1,840,191	7,100	1		1		74,023	103,160			388,129				86,107	Brick, coal, copper, lead, marble, platinum.
1931	1,549,073	4,783	1		1		32,275	57,751			491,456		1		67,933	Brick, coal, copper, lead, marble.
1932	1,307,760	3,865	1		1,454	92	20,284	26,373			19,626			2,981 lbs.	89 / 42,481 / 1,178	Lead. / Brick, coal, marble. / Lead.
1933	1,945,261	6,471	1		13,922	891	18,341	26,016						31,845 lbs.	48,781	Brick, coal, marble, misc. stone.
1934	2,274,275	10,544	1		7,254	580	28,620	50,833			12,115		1	6,102 lbs.	223 / 51,591	Lead. / Brick, coal, gems (diamonds).
1935	2,614,235	17,634	1		9,641	800	37,876	66,654			17,066			3,271 lbs.	800 / 48,779 / 197	Lead. / Coal, brick. / Lead.
1936	3,402,350	18,096	1		31,542	2,902	52,813	91,228			30,777			4,296 lbs.	71,899 / 413	Brick, coal.
1937	3,712,835	18,041	1		18,579	2,248	66,397	107,212			1			7,004 lbs.	77,177	Brick, coal, platinum, misc. stone.
1938	3,724,840	14,569	1		5,152	505	42,679	73,422			6,027				61,081	Brick, coal, lead.
1939	4,167,030	15,411			3,933	409	37,780	64,147			3,300				64,276	Brick, coal, lead, volcanic ash.
1940	4,122,160	16,413			20,643	2,333	34,282	67,164			28,769			11,459 lbs.	573 / 47,447	Lead. / Brick, lead, platinum, volcanic ash.
1941	3,499,300	16,551			11,941	1,409	70,645	130,997			6,088			13,396 lbs.	764 / 69,303	Brick, platinum, volcanic ash. / Lead.
1942	1,731,690	7,887		1	1,854	224	119,596	254,771			17,322		1	10,559 lbs.	708 / 79,538	Lead. / Brick, slate, volcanic ash. / Lead. / Brick, coal, manganese ore, platinum.

Year													
1943	91,210	1,607		624,336	81,164	105,815	236,396		26,426		1	1,429 lbs.	107 Lead. / 97,188 Brick, manganese ore, soapstone.
1944	25,690	1,524	1	440,962	59,530		1		8,492		1	1,560 lbs.	125 Lead. / 187,845 Other minerals.[10]
1945	16,660	8,901		1,619,793	218,672	90,416	194,061		1		1		243,311 Other minerals.[11]
1946	92,330	12,626		2,041,536	330,729	198,264	380,496		3,963		1		118,755 Brick, lead.
1947	165,340	10,496		1,674,000	351,540	189,884	625,537		1				5,317 Other minerals.[12]
1948	231,490	2,038										34,490 tons	65,476 Sand and gravel. / 16,585 Pumicite and stone.
1949	672,770	3,812	39,000	300	59	84,286	160,061		1				29,087 Other minerals.[13]
1950	700,490	3,845	1		110,595	292,125			1			100 lbs.	30,100 Other minerals.[13] / lead.
Totals	$141,774,401	$818,418	$407,121	7,985,204	$1,247,317	2,490,501	$4,375,130	12,640 $15,428	$1,963,966		$427,286		$3,734,326

1 See under "Unapportioned."
2 Includes crushed-rock, rubble, rip-rap, sand and gravel.
3 Includes brick and platinum.
4 Includes brick and soapstone.
5 Includes brick, coal, copper and lead.
6 Includes coal, copper, lead and marble.
7 Includes brick, coal, copper and silica.
8 Brick, coal, manganese ore, platinum.
9 Brick, manganese, soapstone.
10 Brick, clay (pottery), coal, manganese ore, soapstone.
11 Brick, clay (pottery), lead, limestone, miscellaneous stone.
12 Pumice, sand and gravel, and stone.
13 Coal (lignite), sand and gravel, and stone.

Newton Mine.   Location: SW ¼ sec. 28, T. 6 N., R. 10 E., M.D.M., 3½ miles east of Ione on state highway 104.

The Newton mine was located early in 1863 (Jenkins, 1948a, p. 50) during the state's first copper boom, and was worked extensively until the mine was shut down in 1867. The mine was worked intermittently after that date and in 1889 a smelter was constructed. Some of the matte produced at the smelter was shipped as far as Liverpool, England. Production during this period ceased in 1901. As of 1908 total recorded production was 33,000 tons (Aubury, 1908, pp. 50, 224).

The mine was reopened by J. H. Lester early in 1943 and production began in May of that year. The Winston Copper Company took it over and operated it from July 24, 1943, to the end of 1944. From about January, 1945, to the time of closing, August 1, 1947, the mine was operated under contract by the Pacific Mining Company.

The ore is a replacement deposit consisting of a typically massive and fine-grained mixture of pyrite and chalcopyrite. The strike is north-northwest and the dip is from 62° to 70° east, parallel to the schistosity of the amphibole-chlorite schist (Logtown Ridge member) in which it occurs. The vein is from 3 feet to a maximum of 8 feet wide, but generally not over 6 feet. The tenor of the ore probably averages close to 6 percent copper (Heyl and Eric, 1948, p. 55). A series of post-mineral cross faults were encountered by mining operations. One of these on the 700-foot level 15 feet from the station faulted the vein 6 feet eastward. Diamond drill holes run by the U. S. Bureau of Mines indicated that ore extends at least 150 feet further down-dip below the 700 level (Heyl and Eric, 1948, p. 60). Ore reserves in 1943 were estimated by the U. S. Bureau of Mines to be 40,300 tons of 4.5 percent of copper ore.[3]

The mine has been worked on 11 levels from a 700-foot inclined shaft. There are approximately 3,000 feet of underground drifts in addition to many stopes and some crosscuts. The bulk of the mining has been done on and above the 550 level. All the equipment used by Pacific Mining Company was removed from the property in August, 1949.

Production and shipment of ore began in May, 1943, and by the end of the year 50 carloads averaging 60 tons each and running from 8 to 10 percent copper [4] had been shipped to Utah. Total mine production to the end of 1946 is 5,461,132 pounds of copper (Heyl and Eric, 1948, p. 51). The mine has been idle since August 1, 1947.

Gold

Amador County's total recorded production of gold from 1880 to 1950 is $141,774,401 and represents the output from many mines, principally on the Mother Lode, but some on the East Belt. This is about 85 percent of the total mineral production so far recorded for the county. As many of the mines along the Mother Lode were in operation prior to 1880 before production was recorded, that segment of the Lode which passes through Amador County probably has produced gold far in excess of $160,000,000 (Bowen and Crippen, 1948, p. 62).

---

[3] West belt copper-zinc mines, Amador and Calaveras counties, California. U. S. Bur. Mines War Minerals Rept. 103, p. 6, 1943.
[4] Averill, C. V., Unpublished report, December 2, 1943.

From 1837 until 1932, one ounce of gold was worth a fixed price of $20.67. During the depression years of 1932-33, gold was revalued several times at progressively higher values. The average price during this period was $25.56 per ounce. In 1934 it was revalued to its present price of $35.00 per ounce.

Near the beginning of World War II, the War Production Board, a federal agency, issued Order L-208 which went into effect during the fall of 1942. This war-time law restricting gold mining resulted in the closing of nearly all the major gold mines in the country, including those in Amador County.

The Central Eureka mine is the most important active gold mine in the county and is among the few Mother Lode mines which have survived war-time limitations on gold mining and the increased cost of mining operations.

Placer mining has worked most of the gravels of the county. The last of the large bucketline-dredges, operated by the American Dredging Company, ceased operations on the Mokelumne River in the Lancha Plana area in 1923, after having been in continuous operation for 19 years. Smaller placer mining operations have been active intermittently in this county. Volcano was the center for most of the hydraulic mining which took place in the county.

Gold Mineralization. Lode-gold deposits of Amador County are found principally in the Mother Lode vein system or in the East Belt group of veins. The Mother Lode is traceable through the Sierran foothills for approximately 120 miles from the vicinity of Mariposa northwest to Georgetown, El Dorado County. The vein system or vein zone ranges from less than a hundred feet to more than a mile wide. Within this zone are numerous discontinuous or linked veins which may be parallel, convergent at a small angle, or slightly en echelon. Few individual veins can be traced for more than a few thousand feet along the strike but the vein system as a whole is remarkably persistent.

In general, the Mother Lode veins formed in fissures developed within a zone of reverse faulting that strikes north or northwest over much of its length. Repeated movements along the fault fissures facilitated passage of a succession of ascending mineral-bearing solutions which are believed to have had their source in the crystallizing molten rock that formed the granitic core of the Sierra Nevada. Vein matter consists most commonly of milky quartz in which native gold and simple sulfides occur sporadically in shoots, separated by valueless material. In addition to quartz, the most abundant vein mineral, vein matter commonly consists of carbonates, mariposite, chlorite, sericite and sheared wall rock. Native gold occurs alone in various gangue minerals or with sulfides such as galena, sphalerite, arsenopyrite, tetrahedrite, pyrrhotite, and chalcopyrite. Ore shoots generally are short laterally but persist in depth; they tend to pitch steeply either north or south in veins that trend north or northwest and dip steeply east (Knopf, 1929, p. 26). Veins may pinch or swell abruptly and vein ends commonly fray out into veinlets or stringers. Both hanging and footwall sides of the veins may have a zone of adjacent quartz stringers (Knopf, 1929, p. 25). Examination of many of the veins shows that they were repeatedly reopened during formation, giving rise to several generations of vein minerals.

Large bodies of low-grade ore also occur adjacent to Mother Lode veins as wall-rock replacements, particularly in altered greenstone and schist; more rarely, high-grade ore is found in similar environment, as at Carson Hill, Calaveras County. Ore bodies of auriferous schist commonly consist of stockworks where schist masses are laced with gold-bearing veinlets; the schist itself may be partly replaced by ore minerals. Hydrothermally altered greenstone or gray ore, as it is locally called, forms important ore deposits in the southern part of the Mother Lode, notably at the Eagle Shawmut mine in Tuolumne County. Such deposits also occur to a lesser extent in Amador County and in other parts of the Mother Lode.

The Mother Lode vein system cuts a considerable variety of wall rocks, notably the Upper Jurassic Mariposa slate, greenstones and meta-sediments of the Upper Jurassic Amador group, greenstones and schists of the Paleozoic Calaveras group and serpentine. Valuable ore shoots have formed adjacent to all of these wall-rock types but in general, slate appears to have been most favorable for ore deposition and serpentine least favorable (Knopf, 1929, p. 31). Although Mother Lode veins appear to conform roughly to attitudes of the enclosing rocks they actually cut these rocks in dip and less commonly in strike (Fairbanks, 1890, p. 10, and Knopf, 1929, p. 24). Veins commonly are deflected in angle of dip when passing from one type of wall rock to another, the greatest deflection generally occurring at black slate-greenstone contacts (Knopf, 1929, pp. 24-25).

Most of the gold in Amador County is found as native metal, alloyed with a little silver. It may occur in masses easily seen with the naked eye or, more commonly, disseminated in specks of microscopic size. Gold is most commonly associated with quartz, but sulfide minerals containing free gold as a replacement or as fracture filling are in many cases the source of most of the valuable metal. Gold tellurides are found in some parts of the Mother Lode, but are not present in notable quantities in Amador County.

The East Belt district is roughly elongate parallel to the Mother Lode 10 to 15 miles east. It consists of numerous individual veins which do not constitute a closely knit system and do not always have a common trend. Many of the veins have a northerly strike; others are oriented east or northeast. Most of the veins are steeply dipping as in the Mother Lode, but East Belt veins generally are narrower and contain smaller, richer ore bodies. Wall rocks are generally granitic, older metamorphic rocks being much more sparsely distributed than in the Mother Lode belt.

Mines in Amador County that have produced more than $1,000,000 worth of gold.

Name	Total production	Name	Total production
Kennedy	$34,280,000	Zeila	$5,000,000
Argonaut	25,179,160	Fremont-Gover	5,000,000
Keystone	24,000,000	Wildman and Mahoney	5,000,000
Old Eureka	19,000,000	Original Amador	3,500,000
Central Eureka	17,000,000	Oneida	2,500,000
Plymouth Consolidated	13,500,000	Lincoln	2,200,000
South Eureka	5,300,000	South Spring Hill	1,092,472
Bunker Hill	5,154,382	Treasure	1,000,000

Lode Gold Mines

Amador Queen No. 2. Location: sec. 34, 35, T. 6 N., R. 11 E., M.D.M., 2 miles south of Jackson. Ownership: Judge Ralph McGee, Sutter Creek, California.

The chief production of gold from this mine was prior to 1900. By that date the greater part of the underground workings had been developed (Logan, 1934, p. 108). From 1918 (Logan, 1934, p. 60) to 1941 the mine was operated intermittently, with a production of a few thousand dollars nearly every year. The mine was worked intermittently by the Garibaldi Brothers from 1930 to 1941, according to Mr. Frank Garibaldi. In 1948 an unsuccessful attempt was made to treat the tailings.

The Amador Queen No. 2 was worked through a 1200-foot westdriven adit. Drifts were run 1000 feet south and 460 feet north from the end of the adit. Also a winze 900 feet deep was sunk along the dip of the slate with levels at 500 and 700 feet (Logan, 1934, p. 60). In the 1930's a 230-foot winze was sunk about 100 feet from the end of the adit.

Gold occurs with arsenopyrite in quartz stringers in east-dipping Mariposa slate (Logan, 1934, pp. 60-61). The slate is interlayered with schist, some of it being hard and siliceous. In the last ten years of operations, production was chiefly specimen ore.

Amador Star mine. Location: sec. 23, T. 8 N., R. 10 E., M.D.M., 3 miles north of Plymouth. Ownership: Eliza M. Kaiser, 720 West Poplar Street, Stockton, California.

The mine was originally developed through a 422-foot crosscut adit from which some ore was produced prior to 1900. (Logan, 1934, p. 61.) Little mining was done thereafter until 1917 when a 580-foot vertical shaft was sunk approximately 700 feet southwest of the adit portal. Ernest A. Stent continued intermittent development work through the 1920's. More development work by the Amador Star Mining Company was done on the 300 level in 1931.

In 1932, Arthur Hamburger took an option on the property and produced gold during 1933. Under his direction, West America Consolidated Gold Mines, Inc., produced gold from the mine again in 1934. During 1934, a crosscut was extended eastward on the 500 level and other old workings were cleaned out. During the first part of 1935, some drifting was done on the 500 and 800 levels of this mine. The project was abandoned in 1935 as only a few thousand dollars worth of gold was produced and little or no new ore was found, according to Mr. T. Calvert Slater.

The vein, which ranges from 2½ to 9 feet in thickness, occurs in a north-striking half-a-mile wide belt of Mariposa slate. In the wider parts of the vein there are quartz stringers containing considerable pyrite. The small amount of ore milled in the last operation averaged about $6.00 per ton (Logan, 1934, p. 62).

Workings consist of a vertical shaft about 900 feet deep, about 600 feet of drifts on the 300 level, over 600 feet of drifts and a 1000-foot crosscut on the 500 level, and some drifts on the 800 level.

A 75-ton flotation mill equipped with Hardinge ball mills was operated on the property during the early 1930's. Forty men were employed

FIGURE 4.   Headframe of Amador Star mine.

at the mine and mill in this operation, according to Mr. Slater. All of the equipment has been dismantled and removed.

Argonaut Mine. Location: sec. 20, T. 6 N., R. 10 E., M.D.M., one mile northwest of Jackson on highway 49. Ownership: B. Monte Verda and E. C. Taylor, 369 Pine Street, San Francisco, California.

This mine, originally known as the Pioneer, was first worked in 1850. Until 1893, it was a small-scale operation which received little attention. Most of the early work was through an adit on the northern end of the claim (Logan, 1927, p. 153).

In 1893 the Argonaut Mining Company was incorporated. Except for interruptions caused by fires, operations were continuous from this date until 1942. In the spring of 1919 a severe underground fire caused the loss of a year's production as the fire was not brought under control until the lower workings of this mine and the adjoining Kennedy were flooded. On August 27, 1922 a fire on the 3350-foot level caused the loss of 47 lives. This also stopped production for a year (Logan, 1927, p. 154).

From 1923 to October, 1938, tailings produced in the early years of operation were treated by the Amador Metals Reduction Company, according to Mr. S. E. Woodworth, metallurgical engineer who worked the tailings. In 1941, the Argonaut Mining Company constructed a cyanide plant to treat concentrates from the Argonaut mill and those from the Plymouth Consolidated mine.

Although ore in the mine was by no means exhausted, production ceased on March 28, 1942. This was due to the high costs of materials, the scarcity of labor, and wartime limitations on gold mining. Until 1948, however, the upper part of the mine was kept unwatered and in repair in hopes of eventual reopening. In December 1947, the directors of the company recommended dissolution of the Argonaut Mining Company and by February 1948, a majority of the stockholders had given their consent.[5] The mine has since passed into the hands of B. Monte Verda and E. C. Taylor of San Francisco. The machinery is being removed.

The Argonaut vein ranges from 8 to 10 feet in width in the upper workings. At the 4200 level and below, the vein widens to 20 and 30 feet. The vein strikes from N. 10° W. to N. 18° W. At the 290 level the vein dips 40° northeast. From the 470 level to the 4050 level the vein dips 64°, and below the 4050 level the dip ranges from 60° to 63° northeast (Tucker, 1914, p. 19).

The Argonaut vein occupies a fissure apparently opened by a reverse fault (Knopf, 1929, p. 67). To a depth of 290 feet, it is in altered metavolcanic rocks. From the 290-foot level to the 470-foot level the vein cuts a belt of Mariposa slate. At the point where the vein passes into the slate it appears that the thrust fault has a throw of 120 feet, the hanging-wall being thrust upward over the footwall to at least that distance (Tucker, 1914, p. 19). To a depth of 2500 feet the vein is on or near the contact of the Mariposa slate footwall and metavolcanic hanging wall.

A stringer vein which branches into the hanging-wall from the Argonaut vein at a depth of 1200 feet was the subject of a well-known controversy. The Kennedy Extension Mining Company in their famous lawsuit with the Argonaut, contended that this vein belonged to them since it was the older of the two veins. The courts ruled in favor of the Argonaut Mining Company (Logan, 1934, p. 63).

Below the 2500 level the footwall is largely Mariposa slate, and the vein is separated from the hanging-wall of gray-colored schist by a thin band of slate.

The best ore occurred in a ribboned structure of quartz and slate which was within a few feet of the footwall. This ore was free-milling and contained about $2\frac{1}{2}$ percent of sulfides, largely pyrite (Logan, 1934, p. 67). Specimen rock was found in places.   .

The Argonaut mine was developed through a three-compartment 60° inclined shaft 5700 feet deep. The deepest level of the mine was the 6300-foot level which was developed from an inclined winze sunk from the 5550 level 300 feet south of the shaft. The 6300-foot level has a vertical depth of 5570 feet. In the last two years of operations, the principal production was from stopes between the 6150 and 6300 levels. During this same period there was development work underway on these levels and to the south on the 4800 level. There are approximately 8 miles of drifts, crosscuts, and tunnels, 4 miles of raises, and 50 miles of stope floors (Bowen and Crippen, 1948, p. 64).

Before 1936, milling was done in a 60-stamp mill. After stamping, the pulp was concentrated, classified, and the tailings were cyanided. In 1936, ball mills and flotation cells were installed making it possible to raise recovery to 94 percent.[6] A process for elimination of graphite slimes from the flotation pulp by controlled agitation was developed at the Argonaut. This process greatly increased the efficiency of the cyanidation of Mother Lode ores.

[5] Logan, C. A., unpublished report, 1948.
[6] Logan, C. A., unpublished report, 1948.

To December 31, 1942, total mine production was 2,750,000 tons of ore from which $25,179,160.43 was recovered and dividends of $3,789,-750 were paid on an original capitalization of $1,000,000.[7]

Ballard Mine. Location: E$\frac{1}{2}$W$\frac{1}{2}$ sec. 14, T. 8 N., R. 10 E., M.D.M., a quarter of a mile south of the Cosumnes River on highway 49. Ownership: Ballard Mother Lode Mines, Inc., c/o John Ratto, Sutter Creek, California.

Originally worked in the 1870's by the Spanish Mining Company, this gold mine lay idle until the Lopez Mining Company reopened it in 1928 (Logan, 1934, p. 70). The main shaft was reopened and a 10-stamp mill was installed. The operation failed soon afterward and the property was idle until 1933. John Ratto purchased the mine from the Ballard estate in 1931. The property was again active from 1933 to 1937 and from 1941 to 1942. A small amount of work was done in 1947. According to John Ratto, the Ballard Mother Lode Mines Company was incorporated in 1935 with Ratto as the principal stockholder. The mine is now idle.

There are three parallel quartz veins approximately 600 feet apart in Mariposa slate. They strike N. 10° W. and dip 68° east. The ore, which is chiefly free gold associated with pyrite in a ribbon structure of quartz and slate, occurs in a 250-foot ore shoot in the west vein, west of the main shaft, in a 550-foot shoot in the middle vein east of the main shaft and in another smaller 60-foot ore shoot on the middle vein 1000 feet to the south, according to Mr. Ratto.

The main shaft, which was sunk between the west and middle veins, is 285 feet deep with approximately 600 feet of crosscuts connecting both main ore shoots at the 200-foot level. The Ballard shaft, south of the main shaft, is 200 feet deep with 400 feet of crosscuts north of the shaft. Ore exposed in these southern workings has not been completely developed. According to John Ratto, assays have been as high as 180 dollars per ton but the average is around 10 dollars per ton in $35 per ounce gold.

Belden Mine. Location: NW $\frac{1}{4}$ sec. 26, T. 7 N., R. 13 E., M.D.M., about one mile southeast of Buckhorn Lodge. Ownership: Belden Amador Mines, Inc., P.O. Box 28, Fort Wayne, Indiana.

This mine was mentioned in 1867 (Browne, 1868, p. 80) as having been in operation ten years, with a roasting furnace and a five-stamp mill. During this early work a depth of 250 feet was reached.[8]

Little work was done in later years until the present company began operations on September 13, 1936. The mine was shut down in 1938 and re-opened in 1939 by the Belama Corporation, a leasing concern composed of stockholders of the Belden Amador Mines, Inc. The Belden mine was the largest gold producer on the East Belt in Amador County from 1940 until government war-time restrictions closed the mine in November, 1942. The mine was reopened in 1947 with the resumption of ore production in 1948, according to Leon Banks, Superintendent. In 1949, the mine again shut down for lack of water necessary for mining and milling operations. At the present time the mine is inactive pending the outcome of a lawsuit with the Pacific Gas and Electric Company.

[7] Logan, C. A., unpublished report, 1948.
[8] Logan, C. A., unpublished report, 1948.

FIGURE 5.   Belden mine. Mill in right background, hoist house on left.

The gold occurs in stringer quartz veins in granodiorite. The main vein strikes N. 15° W. and dips east. There are three en echelon ore shoots that dip steeply to the east. The gold content in the ore is 50 percent free gold and 50 percent auriferous pyrite with a little pyrrhotite, according to Mr. Banks. Minor amounts of galena and sphalerite are also present. The ore bodies are faulted to the west toward the south with a displacement amounting to 20 to 25 feet along three major faults. As a result of hand sorting the ore before milling, assays of the mill feed averaged $56 per ton during 1948. The average gold value prior to this date was $41 to $42 per ton, according to Mr. Banks.

The 160-, 250-, 300-, and 350-foot levels were driven off a 400-foot, two-compartment, vertical shaft. Since 1936, the company has worked the mine from the 200-foot to the 400-foot level and has stoped a length of about 400 feet along the strike. There are several thousand feet of drifts in the mine, some of them being as much as 500 feet in length. The ore is hoisted from the mine in 1500-pound buckets.

Ore is beneficiated in the company mill, the capacity of which is twenty tons per day with a 92 to 93 percent recovery. The ore is washed, hand sorted, and crushed by a primary jaw crusher and a secondary ball mill. The fines are pumped to a Diester rubber-covered table and concentrated into free gold, pyrite concentrates and middlings. The free gold is amalgamated in a Denver amalgam barrel, retorted, and sold as bullion to the mint in San Francisco. The middlings are run through a corduroy-lined launder, and then to a classifier from which the oversize is returned to the ball mill while the fines are sent to a four-cell Denver flotation machine. The froth concentrate, along with the pyrite concentrate, is filtered, dewatered, and shipped to the Empire Star mine in Grass Valley for further treatment.

The mine has produced gold worth $400,000 since 1936, according to figures furnished by Mr. Banks. When in full production, 25 men operate the mine and mill, producing 40 tons per day.

Black Prince Mine. Location: SE ¼ sec. 27, T. 7 N., R. 13 E., M.D.M. about 1½ miles southeast of Buckhorn Lodge. Ownership: This property consists of three unpatented mining claims, the Coeur Leonie, the Coeur Leonie fraction, and the Aurora, owned by a partnership ·omposed of P. M. Wedell and W. R. Schwickerth who resides at the nine.

This property was operated by Jack Howald in the early 1930's and produced a total of $50,000 worth of $35 per ounce gold during this time, according to Mr. Schwickerth. In 1936, the property was sold to the present owners and since that time, small amounts of work have been done intermittently.

Free gold occurs with pyrite and smaller amounts of chalcopyrite in a series of quartz veins which strike approximately N. 8° W. and dip steeply to the east. The country rock is granodiorite. The Black Prince vein, which is the main vein, is about 3 feet wide. The Black Fox vein is 550 feet west of the Black Prince vein while the Aurora vein is 550 feet east of the Black Prince vein. Assays from ore shoots of the Black Prince vein average $90 per ton in $35 per ounce gold, according to Mr. Schwickerth.

The main workings consist of three adits which have been driven along the Black Prince vein in a general direction of N. 8° W. These adits are each separated by a vertical distance of about 100 feet. The lower adit has been driven 600 feet, the middle adit, 450 feet, and the upper adit, 225 feet. Mining and development work consists of crosscutting, raising, and stoping off of these levels. A 720-foot adit has been driven on the Aurora vein and an 85-foot adit has been driven on the Black Fox vein. There is no mining equipment on the property.

Black Wonder Mine. Location: SE ¼ sec. 5, T. 6 N., R. 12 E., M.D.M., two miles southwest of Pine Grove just east of state highway 88. Ownership: George B. Taves and Florence J. Taves, Pine Grove, California.

Work started at this mine in 1931 under a partnership consisting of Bert Caldwell, Nick Contini and Dean Wiley. In 1933 and 1934 it was sub-leased to Weise and Holmes. During 1933, 800 pounds of ore containing $173 worth of $20.67 per ounce gold was shipped to the smelter of Selby, California, according to Mr. Caldwell. The mine was shut down in 1935 and has been idle since that time.

Free gold associated with pyrite and small amounts of chalcopyrite and some galena occurs in a 3- to 4-foot-wide quartz vein. The vein strikes northwest and dips 60°-70° southwest. The country rock consists of blocky greenstone.

Up to 1933, the inclined shaft was 90 feet deep. In 1934, the shaft was retimbered and sunk an additional 30 feet. Also at this time a 30-foot drift was driven in a northwesterly direction off the 45-foot level. A 9-inch vein of galena was encountered near the end of this drift, according to Mr. Caldwell.

Ore was treated in a mill equipped with a jaw crusher, an air-operated five-stamp mill and amalgamation plates. Four to six men operated the mine and mill.

Bunker Hill Mine.   Location: sec. 25, 36, T. 7 N., R. 10 E., M.D.M., ¼ mile north of Amador City. Ownership: Bunker Hill Mining Company, c/o D. C. Crandall, North Bend, Washington.

Originally known as the Rancheria mine, the Bunker Hill was first worked in 1853. During the 1860's and 1870's, ore yielding from $50 to $75 per ton was being produced (Logan, 1934, p. 72). An inclined depth of 800 feet was reached in 1888. In 1893, after having been idle for several years, the mine was renamed the South Mayflower and reopened.

In 1899, the Bunker Hill Consolidated Mining Company was organized and operated the mine steadily until 1922. During this period 887,585 tons of ore was produced which yielded $3,834,550. Since 1922 the mine has been idle except for a small amount of work done in 1925. A total of $5,154,382 in gold was produced (Logan, 1934, p. 72). The company paid dividends of about $1,000,000 on an original capitalization of $200,000.

The principal vein is the Bunker Hill vein, or the hanging-wall vein. Other veins, including the "gouge vein" and the Last Chance vein, lie in the footwall (Knopf, 1929, p. 55). From the surface to the 200 level the Bunker Hill vein occurs along a greenstone-Mariposa slate contact, the greenstone lying to the east (Tucker, 1914, p. 20). The vein is in slate from the 200 to the 1400 levels, and below the 1400 level the vein crosses interlayered bands of slate and greenstone. The vein strikes northwest, dips to the northeast, and averages five and a half feet in width. Gold associated with pyrite and some arsenopyrite occurs in quartz veins and greenstone ("gray ore"). Some mineralized black slate, heavily charged with both pyrite and arsenopyrite, that was in the footwall at the north end of the 2200-foot level proved to be ore (Knopf, 1929, p. 55). West of the Bunker Hill vein on the 1400 level a body of auriferous greenstone ("gray ore") was developed.

The mine was developed by a two-compartment, 2800-foot inclined shaft sunk on an angle averaging 58°. An inclined winze was sunk from the 2800 level to a depth of 3440 feet. The last work was done on the 3200 and 3400 levels: Ore was mined by square-set stopes and later filled with waste. Shrinkage stopes were employed in the auriferous greenstone.

Ore was milled in a 40-stamp mill. Milling was accomplished by primary crushing, stamps, amalgamation, and concentration on Deister tables. Sands were reground in a ball mill and the entire pulp was then concentrated on 24 Frue vanners.

Central Eureka (Including Old Eureka) Mine.   Location: SW ¼ sec. 8, T. 6 N., R. 11 E., M.D.M., at the town of Sutter Creek. Ownership: Central Eureka Mining Company, 2210 Russ Building, San Francisco, California.

Previous to 1924, the Central Eureka mine and the Old Eureka mine were operated as two independent gold mines. In 1924 the Central Eureka Mining Company, which had been operating the Central Eureka mine for 28 years, acquired title to the Old Eureka mine and the two mines were consolidated (Logan, 1934, p. 103).

The Central Eureka mine was originally located as the Summit mine in 1855. It worked continuously from that date until 1875. The mine was reopened in 1895 (Storms, 1900, p. 64) when the Central Eureka

Mining Company began operations. Profitable exploitation continued until 1907 when the mine was shut down for one year, reopening again in 1908. During 1924, the company purchased holdings of the Old Eureka mine but it was not until 1930 and thereafter that the production of ore had shifted from the Central Eureka workings to the Old Eureka workings (Logan, 1934, p. 75).

The Old Eureka mine was opened in 1852. In 1859, Alvinza Hayward consolidated an adjoining claim with the Old Eureka and by 1867 56 stamps were crushing ore. In March, 1869, the mine was purchased by the Amador Mining Company for $750,000. (Logan, 1934, p. 101.) The new operators continued mining until 1881. The mine was idle from 1881 to 1916 when dewatering and mining operations were resumed; it was again shut down in 1920. Some work was done in 1924 at the time the present operators purchased the mine but it was not until 1930 that the lower workings of the Central Eureka mine were abandoned in favor of the ore bodies of the Old Eureka mine. All of the ore produced since 1930 by the Central Eureka Mining Company has been from the Old Eureka workings. Mining operations were suspended in 1942 but the mine was kept in working order during World War II. Mining was resumed in 1946, but little ore was produced until 1948.

In 1951, 39,440 tons of ore yielded a little more than $500,000 in gold and silver, according to Morton A. Ralls, superintendent. Total production for the Central Eureka mine, which includes the Old Eureka and Central Eureka, to the end of 1951 is approximately $36,000,000.

Mineralized zones in the Central Eureka and Old Eureka mines consists chiefly of quartz, quartz-ankerite-albite rock, ribboned structures of quartz and slate and hydrothermally altered greenstone or "gray

FIGURE 6.   Shaft of Central Eureka mine.

FIGURE 7.   Central Eureka mill and cyanide plant. Flag in upper righthand corner
shows location of Central Eureka shaft.

ore.'' Gold alloyed with silver may be in a free-milling state in any of these rocks or it may be disseminated in pyrite where fine crushing is necessary to free it from the host minerals. Pyrite and arsenopyrite are common and sphalerite, chalcopyrite and galena are present in minor amounts, according to Mr. Ralls.

The veins strike essentially north, dip steeply eastward and, although they pinch and swell, have an average width of about 8 feet. In some places the width may reach 20 feet. Wall rocks may be Mariposa slate, Logtown Ridge greenstone, or Cosumnes graywacke, all of Jurassic age. Slates commonly form the footwall along major faults.

The most pronounced structural feature in the Old Eureka mine is the Wolverine reverse fault, which on an average strikes north 20° west and dips 65° east. (Norman, 1939.) In the upper 1700 feet of the mine, faulting has occurred along the contact of Mariposa slate and Logtown Ridge greenstone. Below the 1700-foot level, the footwall remains slate but the hanging wall becomes a series of dense and slaty greenstones. (Logan, 1934, p. 103.) Between the 2500 and 3000 levels, the measurable displacement was observed to be more than 300 feet. Rich ore bodies occur along the Wolverine fault as irregular shattered quartz lenses within gouge.

Workings consist of two inclined shafts, the Central shaft which is 4855 feet deep, and the Old Eureka shaft which is 3500 feet deep. The Central shaft is 1800 feet south of the Old Eureka shaft. Each is a two-compartment shaft with an average incline of 65° east. The lower levels off the Old Eureka shaft are worked through an 11- by 5-foot inclined winze sunk from the 3500 to 4150 level. The main haulage way for the mine is the 3500 level. All of the ore is hoisted up the Central shaft while the haulage of men, waste, timber, supplies, and occasional bailing of water is confined to the Old Eureka shaft.

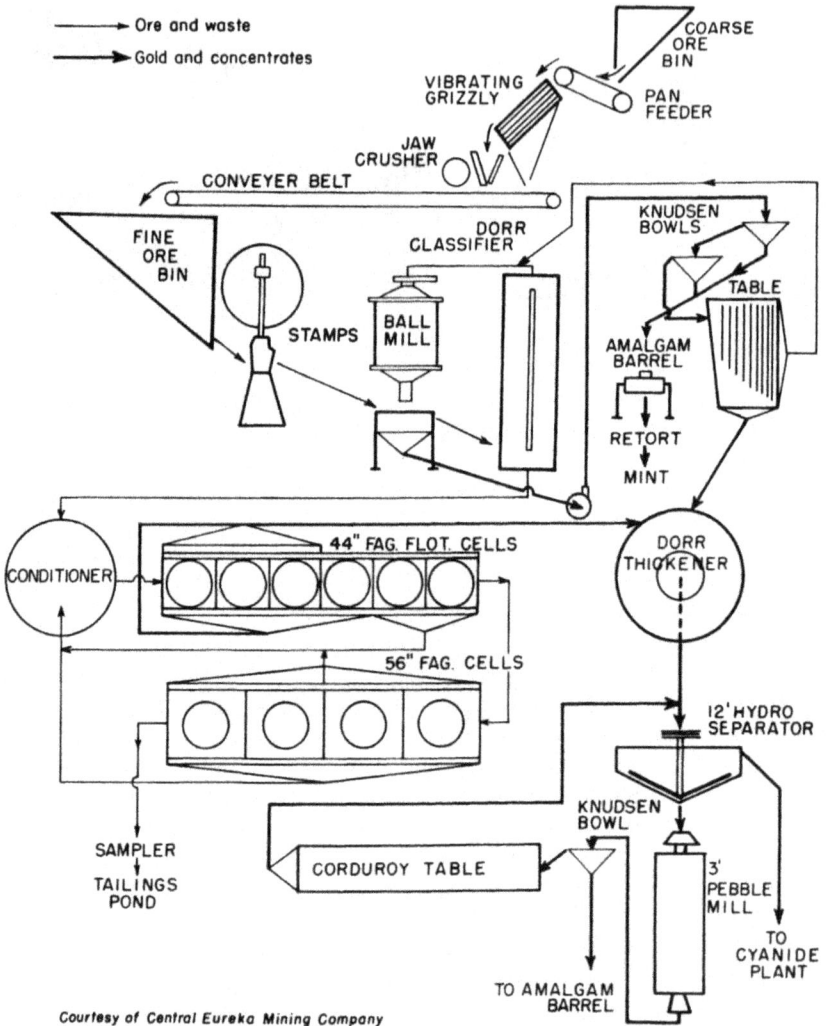

FIGURE 8.   Flow sheet, Central Eureka mine, courtesy Central Eureka Mining Company.

The cut-and-fill method of stoping is used. Since 1946, sand has been used to fill the stoped-out areas, but prior to World War II, waste was used, according to Mr. Ralls. Swedish Copco Atlas pneumatic drills with detachable bits and tungsten carbide inserts are used for drilling. Ore from the stopes is loaded into mine cars, trammed to transfer chutes, and hoisted up the 3500 winze to the 3500 level. Electric battery locomotives transport the ore-laden mine cars 3000 feet south to the Central shaft where the ore is hoisted to the surface in 3-ton skips.

As there is considerable gouge ranging from 1 to 6 feet in thickness on both the hanging and footwall on many of the levels, stopes must be timbered close to the face.

Since 1948, mining has been done on the 2900, 3100, 3400, 3500, 3600, 3700, 3800, 3900, and 4150 levels. The principal work being done at the present time is north of the Old Eureka shaft on the 3700, 3800, 3900, and 4150 levels through the 3500 winze.

Between 1935 and 1938, the Central Eureka Mining Company ran exploration drifts north from the 2100 and 2500 levels of the Old Eureka workings into the Lincoln Gold Mining Company's property. A total of 2467 feet of drifting and crosscutting was completed on both levels, but no commercial ore was located, according to T. Calvert Slater, former superintendent.

Ore is run through a jaw crusher, 30-stamp mill, jig, classifier, conditioner, and flotation cells. The gold and concentrate from the jig is delivered to two Knudsen bowls, an amalgam barrel, retorted, and sent to the U. S. Mint in San Francisco. The gold and concentrate slimes from the Knudsen bowls are tabled and sent to a Dorr thickener. Slimes from the flotation cells are also sent to the Dorr thickener. The gold and concentrates from the thickener are delivered into a 12-foot hydro-separator. The products from this separation either go to the cyanide plant or are reground in a "pebble mill" and sent to another Knudsen bowl. The fines from the bowl are delivered to the amalgam barrel while the slimes are run over a corduroy table and recirculated through the hydro-separator. Mill recovery averages about 95 percent. Mill capacity is 9 tons per hour. Approximately 70 percent of the gold is recovered by amalgamation and 30 percent is recovered from the cyanide concentrates. Ratio of concentrates sent to the cyanide plant is from 30-40 to one.

The mill discharge of sand is delivered into settling tanks where one pound of aluminum sulfate is added to each ton of discharge to aid in settling. Thirty percent of the discharge is discarded on the tailings pond, while the remaining 70 percent of the sand is mixed with water to give a 65 percent solid—35 percent liquid mixture. This mixture is piped down the Old Eureka shaft and used as fill. It is possible to deliver up to 30 tons of sand per hour for use as filling, according to Morton A. Ralls.

Three shifts of men per day operate the mine on a six-day work week. Mining is done during two shifts while repair work is done on the third shift. Miners are paid on a contract basis and are paid by the linear foot. The total payroll includes 90 to 125 men.

Contini (Mikado) Mine. Location: W ½ sec. 9, T. 6 N., R. 12 E., M.D.M., 1.8 miles southwest of Pine Grove. Ownership: Bert Contini of Jackson owns two unpatented mining claims, the IXL and the Three Horsemen.

In 1889-1890, this mine was worked in conjunction with the Wheeling mine, one mile to the east, according to Nick Contini. In 1932 the claims were relocated by Nick Contini and were worked from 1935 to 1939 and again in 1950.

Coarse free gold occurs in quartz with pyrite and small amounts of chalcopyrite and galena. Secondary blue and brown opal are common on this property. The vein varies from 6 to 12 feet in width, strikes northeast, and dips 60° to 70° southeast. Country rock consists of blocky greenstone cut by granitic dikes.

The workings on the IXL claim are a 250-foot southeast-trending tunnel, a 60° to 70° inclined, now-caved winze and approximately 200 feet of crosscuts. The Three Horsemen claim has a 20-foot adit about 200 feet east of the IXL adit portal. There is no equipment on the property.

In the period of 1935 to 1939, about $6000 in gold was produced. Small amounts of ore were produced in 1950 and during the winter of 1951-52 when about 20 tons of ore were shipped to Pine Grove for milling.

Defender Mine. Location: SE¼SW¼ sec. 29, T. 7 N., R. 13 E., M.D.M., half a mile southeast of Pioneer Station. Ownership: West Point Consolidated Mines Company, c/o E. H. Outerbridge, 250 Park Avenue, New York, New York.

The mine was worked intermittently under lease from the 1900's to 1938. Between $20,000 and $30,000 of gold was produced during the 1930's, according to R. Moar, former miner. The mine has been inactive since 1938.

Free gold and pyrite associated with some galena, sphalerite and chalcopyrite occur in a 20-inch-wide quartz vein. The vein strikes N. 15° W. and dips 85° southwest (Tucker, 1914, p. 27). Country rock is granodiorite. The last ore produced in 1938 assayed from $7.00 to $9.00 per ton in gold, according to Mr. Moar.

The mine is exploited by a 480-foot inclined shaft. On the 200 level, a 300-foot north drift and a 150-foot south drift were driven. Most of the ore produced was between the 200 level and the surface. There is no equipment on the property; the shaft is caved. During the last year of operation, 1938, a 75-ton milling plant was completed. Approximately 35 men were employed at the mine and mill, according to Mr. Moar.

Elkhorn Mine. Location: SW¼SE¼ sec. 29, T. 7 N., R. 13 E., M.D.M., a quarter of a mile southeast of Pioneer Station. Ownership: This property is a patented mining claim owned by Mary and Sadie R. Grillo of Volcano, California.

Two caved shafts, one 100 feet north and the other 300 feet south of the Elkhorn mine shaft, are part of the old workings that were actively mined during the early 1890's. Edward Schaefer of Pioneer Station leased the property in 1939 and operated it until government order L-208 closed the mine in 1942. The property was leased to the present operators, a partnership consisting of V. F. Pierce and W. H. Riddle of Sacramento, and F. C. Hamburg of Loomis in January of 1950, according to Mr. Pierce. At the present time the partnership is sinking a new shaft. The property is worked intermittently.

Free gold occurs in a 2½-foot wide north-striking quartz vein and stringers in granodiorite. Associated with the gold is pyrite and some chalcopyrite, sphalerite, bornite, and arsenopyrite.

The mine has a two-compartment vertical shaft, 85 feet deep, with a 20-foot drift north and a 35-foot drift south, both off the 75-foot level. Ore is brought to the surface by a three cubic-foot bucket, the hoist of which is powered by a 1928 Chevrolet engine. A model 315 Worthington gasoline compressor supplies compressed air to operate the pneumatic rock drills while a 450 g.p.m. pump is used to dewater the mine. The vertical Haywire shaft, now caved, is 300 feet south of the Elkhorn shaft; the former is 175 feet deep, with two levels at 125 and 175-foot depths.

Fort Ann (Acme, Robinson) Mine. Location: E ½ sec. 2, T. 7 N., R. 12 E., M.D.M., 3½ miles north of Volcano. Ownership: William L. Metcalf, Box 32, Volcano, California.

This mine was worked prior to 1895 (Crawford, 1896, p. 66) and in 1895 there was a 10-stamp mill on the property. Between the late 1890's and 1935 the mine was largely inactive. In 1934 J. C. Nimmo worked the mine tailings for gold. In 1935, the Fort Ann Mining Company, which was organized by J. C. Nimmo and F. W. Kent, both of Los Angeles, and William Anderson of San Diego, reopened the mine. Mining operations continued steadily from 1935 until 1941, according to Mr. George Garland, former miner. In the fall of 1939, the company had completed erection of a 75-ton-daily-capacity milling plant on the property. In September, 1944, W. L. Metcalf and G. C. Rubke, both of Volcano, leased the mine. There has been no production of ore since 1941 and only maintenance work has been done at the mine since that date.

The ore occurs as free gold with accompanying pyrite in a quartz vein and stringers. Country rock is green schist and slate. A 300-foot nearly vertical shaft is now caved.

During the 1930's, a west adit was driven several hundred feet into the side hill, according to Mr. Lee Gardner, former miner. From this adit, other crosscuts and drifts were driven.

Ore was hand trammed from the mine to the 75-ton mill which is located about 200 feet southeast of the adit portal. The ore was crushed in a jaw crusher and ball mill, conveyed to a trommel, Dorr classifier, and two Wilfley tables. The concentrates from the tables were sent to the amalgam barrel while the tailings were deposited in a tailings pile. The electrically powered mill remains on the property and the adit is kept open.

Fremont-Gover Mines. Location: sec. 25, T. 7 N., R. 12 E., M.D.M., one mile east of Drytown. Ownership: Fremont Gover Company, c/o C. B. Braun, Route 3, Box 306, Albany, Oregon.

The Loyal Lode mine, now part of the Fremont Gover, is known to have been in operation prior to 1867. In 1872 the Fremont and Gover Company was formed, and during the 1880's and 1890's worked principally on the Gover claim (Logan, 1934, p. 82). In the meantime the Loyal Lode was reopened and produced ore yielding between $7 and $8 a ton. Production from the Gover varied from $50,000 to $70,000 per year in the late 1880's and early 1890's.

FIGURE 9.   Fremont mine. Scale is shown by man standing
in front of headframe.

In 1900 the Fremont shaft was started and in 1903 a new 40-stamp mill was erected. Operations continued until late in 1918 (Knopf, 1929, p. 52). From 1920 to 1923, Metals Exploration Company prospected and produced some ore from the property. A small amount of work was done in 1925 by Fremont-Gover Mines Company, organized by former employees. Shortly after, Black Hills Fremont Mines Company did a little work in an adit and in a 100-foot winze.

Early in 1937 the property was reopened by Amador Mother Lode Mining Company. In September of 1937 the property was purchased by the Fremont Gover Company who operated it until April of 1940. In this last operation there was some production. The Gover shaft was reopened to the 865 level and the Fremont to the 770 level. Several thousand feet of new drifts, crosscuts, and raises were driven.

Total production of the property has been in excess of $5,000,000 (Bowen, 1948, p. 65). Dividends of $316,000 were paid.

The gold-quartz vein on the Fremont-Gover property strikes N. 25° W. and dips 50° east. The average thickness of the vein is 6 feet (Tucker, 1916, p. 28). Gold was produced from slate-quartz ore, pockets in quartz stringers extending into the greenstone, and auriferous greenstone or "gray ore." Arsenopyrite, as well as pyrite, is commonly present.

At the Gover mine, the vein occupies a fault-fissure having an observed displacement of 375 feet. From the surface to the 170-foot depth, the vein is in slates while from that depth to the 600 level the vein has a greenstone hanging-wall and a slate footwall. Below the 600 level, the vein is in greenstone. The vein is narrow in the slates, but widens in depth, 50 and 60 feet being commonly reached (Knopf, 1929, p. 54).

Below the 900-foot level and extending to the 2300 level in both the Gover and Fremont claims was a large body of "gray ore." The "gray ore" occurred in a zone up to 70 feet wide. (Logan, 1934, p. 84.) Several other lenses of "gray ore" were mined between the 800 and 2500 levels. Quartz-slate and pocket veins were developed in the upper levels of the Fremont shaft.

The Fremont shaft is 2950 feet deep on a 51° incline. Levels are 200 feet apart. The main Gover shaft, which is 1430 feet north of the Fremont, is 1500 feet deep on an incline. Drifting was extensive from both shafts. The 1500 level of the Gover connected with the 1350-foot level of the Fremont. Mining was done partly by square setting and partly with stulls, all stopes being filled. (Logan, 1934, p. 84.)

The mill contained 40 stamps and had a capacity of 200 tons per day. Concentration was on Frue vanners. From 1914 to 1920 the tailings were worked on a royalty basis by the California Slimes Concentrating Company. Tailings were treated by regrinding, concentration, and cyanidation.

Hageman Mine. Location: NE. ¼ sec. 33, T. 7 N., R. 13 E., M.D.M., 500 feet east of the West Point Power House. Ownership: Four unpatented claims owned by Frank E. Cotie and Vernie Hoffschneider and leased by a partnership consisting of E. V. Grant, R. R. Brown, and E. C. Anderson.

The Hageman adit was driven in the early 1920's and some high grade ore was taken from the Hageman vein. (Logan, 1922, p. 7.) In 1947 the Sunny Boy vein was uncovered during the construction of the new West Point road. The property was leased soon afterward by the Dowdel Brothers, who sank a 50-foot shaft on the Sunny Boy vein and produced over $20,000 to 1951. The present operators leased the property early in 1952 and are working intermittently to extend the Sunny Boy adit west to encounter the Sunny Boy vein, according to Mr. Grant.

The Sunny Boy vein strikes north and dips 52° west. It ranges from a few inches up to 3 feet in width. The gold occurs in pockets and is associated with pyrite and small amounts of galena.

Workings consist of the 150-foot northward-extending Hageman adit, the Sunny Boy adit 200 feet to the west, which extends 150 feet in a westerly direction toward the Sunny Boy vein and the 50-foot inclined shaft about 150 feet north of the Sunny Boy portal. At the 50-foot level there is a 60-foot drift south.

Italian (Black Hills) Mine. Location: SW¼ sec. 24, T. 7 N., R. 10 E., M.D.M., about one mile northeast of Drytown. Ownership: Black Hill Mining Company, c/o William Tam, Jackson, California.

The mine began producing gold in the early 1860's (Logan, 1934, p. 85), after which it was idle until operations were resumed for a short period during 1890. It was reactivated in 1932 and remained in operation until 1940. Since that time, no production has been recorded.

Free gold, which occurs in quartz between a slate footwall and a greenstone hanging wall, has been recovered from several thousand feet of crosscuts, winzes, drifts, and stopes.

A five-stamp mill was used during the productive years of 1932 to 1940. At the present time, there is considerable mining equipment remaining on the property, including grinding equipment, concentrating tables, hoists, compressors, several hundred feet of track and some ore cars.

Between the years 1932 and 1940, approximately $140,000 worth of gold was produced, according to figures provided by the owner.

Jumbo Mine. Location: SW ¼ sec. 26, T. 7 N., R. 13 E., M.D.M., about 2 miles southeast of Buckhorn Lodge. Ownership: Four unpatented mining claims are owned by a partnership consisting of Behrend Doscher, George Viscovich, and Steve Miloservich.

These claims were located by Joe Lawrence of West Point during 1950. The present owners purchased the claims in 1951, according to Mr. Doscher.

There is a north-south belt of mineralization approximately 6 feet wide in granodiorite. Within this zone are 1- to 3-inch quartz stringers and quartz pockets. The ore consists of free gold with quartz, some pyrite, and smaller amounts of chalcopyrite and malachite. The vein crops out 3000 feet along the strike north of the river. Near the river the vein is nearly vertical but toward the north it dips to the west.

Parts of the vein in the second adit assay $12 per ton in gold, $12 per ton in the open cut, and $121 per ton in the main zone on the lower level. Hanging-wall samples in the granite vary between $14 and $28 per ton in gold, according to information received from Mr. Doscher.

Development work consists of a 20-foot north-trending adit 50 feet above the river, a second 65-foot, northeast-trending adit 125 feet above the first adit on the same level as the Pacific Gas and Electric Company's road, and a 20-foot open cut 65 feet above and to the east of the second adit.

There is a 60° inclined tramway between the second and first adits. An air-operated hoist with a ¾-ton skip utilizing a 25-foot wooden headframe hauls the ore up the tramway from the lower adit to the level of the road where the mill is located. The ore is crushed to minus ½-inch by a jaw crusher and conveyed by conveyor belt to a ball mill. Free gold is recovered in a Knudsen bowl while the fines are sent to the classifier and flotation cells. The electric power is supplied by Pacific Gas and Electric Company.

Kennedy Mine. Location: SW¼SW¼, sec. 16, T. 6 N., R. 11 E., M.D.M., one mile north of Jackson. Ownership: Mark and Frances Eudey, Martell, California.

The original Kennedy claim was located in 1856 and in 1860 several properties were consolidated to form the Kennedy mine. These properties were patented in 1872. Little work was done until the Kennedy Mining and Milling Company purchased the mine in 1885.

A 40-stamp mill was erected in 1886. During the 1890's approximately 36,000 tons of ore per year was being produced (Logan, 1934, pp. 86-87). In 1900 the present east vertical shaft was started. From that date on the Kennedy was one of the largest producers in the Mother Lode. Between 140,000 and 170,000 tons of ore per year was produced in the 12 years prior to 1916 (Logan, 1934, p. 87). A fire in 1919 in the Argonaut mine spread to the Kennedy workings through tunnels which connected the two mines. Production was halted as the fire was not brought under control until the lower workings of both mines were flooded.

FIGURE 10.   Kennedy mine. To right is partially dismantled stamp mill.

As the mine increased in depth and mining costs increased, it was necessary to mine more selectively so that from the middle 1920's until 1942 the tonnage of ore produced decreased, according to Mark Eudey, owner and former superintendent. On September 7, 1928, a disastrous fire destroyed all of the surface plant except for the mill and main office. The surface plant was rebuilt in 1929. In 1935 a 1500-ton cyanide plant was erected south of the mine to re-treat accumulated tailings from the mill. This was in operation until 1939.

From 1929 to 1941 the chief production was between the 4650 and 5900 levels, the inclined winze having been started in 1929.

In August, 1941, increasing costs caused all work below the 4650 level to be discontinued. According to Mr. Eudey, from this time until the mine closed, preparations were underway to mine known ore from that level and above.

The mine was closed in November, 1942, in compliance with U. S. Government Order L-208. Except for the steel headframe, all of the surface plant including the buildings have since been sold.

In 1948-49, the tailings were worked intermittently by Frank Fuller, Jr. of Jackson and from 1949-50 by Michael Hagel of Sacramento. Hagel's project was not successful. In September, 1950, the Kennedy Mining and Milling Company went out of existence, thus ending one of the best known mining operations in California. The gold lodes were, however, by no means exhausted and further mining merely awaits development of more favorable economic conditions.

Total production of the Kennedy mine with the price of gold calculated to fit various price changes is approximately $34,280,000 (Bowen and Crippen, 1948, p. 64). Five and one-half million dollars in dividends were paid on an original capitalization of $100,000.

FIGURE 11.   Three of the tailings wheels at the Kennedy
mine. Mine shows through spokes of wheel to right. *Photo
by Mary R. Hill.*

The main Kennedy vein strikes N. 20° W. and dips 70° E. The ore is free gold in quartz with auriferous pyrite and minor amounts of galena. The ore bodies of this mine are not continuous either longitudinally or in depth. The thickness of ore mined usually ranged from 8 to 15 feet. Generally, the best ore is in ribbon rock of hard white quartz containing numerous ribbons of finely ground slate or fine-grained pyrite and galena, according to Mr. Eudey, and to a depth of 4800 feet the ore remained at a fairly constant gold value; however, at greater depths the ore occurred as irregular masses. Little specimen ore was recovered.

Wallrock consists of Logtown Ridge greenstone and Mariposa slate. Below the 3150 level the vein is wholly in slate as the narrow greenstone mass near the vein in the upper levels pinches out at this depth (Logan, 1927, p. 168). An east vein, which was encountered between the 1700 and 2200-foot levels and extends to the bottom levels, has been formed by the breaking away of a large horse of the hanging wall slate. At the widest part of the horse, this vein is 150 feet from the main vein. This vein joins the main vein both north and south of the shaft.

The Kennedy was developed through a three-compartment vertical shaft 4764 feet deep. Work at the lowest levels was done through an inclined winze off the 4650-foot level. The greatest vertical depth reached was 5912 feet, making the Kennedy the deepest mine in North America. There are approximately 50 miles of underground workings, according to Mr. Eudey.

Prior to 1931, milling was done with a 100-stamp mill. At that time a ball mill and flotation cells were installed. The stamp mill was then used for primary crushing. From 1936 until the mine closed, an average of

3000 tons of ore per month were milled. Total mining and milling costs averaged 10 dollars per ton. The last full crew at the mine and mill was 125 men.

For twenty years prior to 1934, tailings were stored behind a dam south of the shaft. They were transported to the dam by 2000 feet of flume. Tailing-elevator wheels 56 feet in diameter were used. Pictures of these large wheels often appear in publicity and travel folders of the Mother Lode.

Keystone Mine. Location: sec. 36, T. 7 N., R. 10 E., M.D.M., within the city limits of Amador City. Ownership: Keystone Mining Company, c/o John D. Culbert, Amador City, California.

One of the most profitable mines in the Mother Lode was the Keystone which produced $24,000,000 (Bowen and Crippen, 1948, p. 65). The Keystone mine was formed from several claims originally located in 1851. During the early period of operations much high-grade ore was produced so that dividends as high as $550 per share per month were declared (Knopf, 1929, p. 58).

By 1888 most of the high-grade ore had been exhausted. Mining operations continued until the mine closed in 1919. From 1911 to 1919, the Keystone Mining Company operated the mine (Logan, 1934, p. 92) and in 1920 the company gained control of the South Spring Hill mine adjoining the Keystone property on the east side.

In 1933 the Keystone was reopened by Keystone Mines Syndicate. Production was underway by 1935 and continued until the fall of 1942 when the mine was closed down by governmental order L-208. During this last period of operation about $1,000,000 was produced, according to T. S. O'Brien. Since that date the mine has been idle except for a small amount of work done in 1952 in the Wonder mine which is just east of the main shaft.

The main Keystone or contact vein varies from 12 to 200 feet in width and occurs on the contact between slate and greenstone, the former being the footwall and the latter the hanging wall (Tucker, 1914, p. 35). The vein dips 35 to 65 degrees east. East of this vein is the Spring Hill or East vein. Ore, which varied greatly in value, occurred both in the veins and quartz stringers branching off the veins. Ore was free gold in quartz associated with pyrite and small amounts of arsenopyrite and stibnite. Some of the sulfides were auriferous.

The mine was worked through the main or Patton shaft, which is 2680 feet deep on a 52 degree incline and is located between the Spring Hill and Keystone veins. Production was from the 900 to the 2100 levels, although the principal production was from the 1400 to the 2100 levels, according to Mr. O'Brien. Crosscuts run both to the east and to the west. Below the 2100 level there was very little production. Ore was first crushed at the Keystone and then trucked to the Original Amador mine where it was milled and cyanided. Just prior to the closing of the mine in 1942, about 300 tons of ore per day were milled.

About 100 men were employed at the mine and mill, according to Mr. O'Brien.

Lincoln Consolidated Mines. Location: secs. 6, 7, 8, T. 6 N., R. 11 E., M.D.M., in Sutter Creek and extending north for 1 mile along the Mother Lode. Ownership: Lincoln Gold Mining Company, 807 Lonsdale Building, Duluth, Minnesota.

Lincoln Consolidated Mines include the Lincoln, Wildman, and Mahoney mines, which were operated separately until 1906 when the Lincoln Consolidated Mining Company gained control of the three properties. The Lincoln mine is the farthest north, the Mahoney is south of the Lincoln, and the Wildman, the most southern adjacent to the town of Sutter Creek. All three mines were originally located about 1851 and worked during the 1860's.

In the 1860's the Lincoln was worked through two shafts and produced 3500 tons of ore per year (Logan, 1934, p. 95). It lay idle until 1898 when it was reopened by the Lincoln Gold Mine Development Company (Storms, 1900, p. 75). Old workings were cleaned out and there was some production for several years.

The Wildman, idle since 1867, was reopened in 1887 and was in operation until 1906. The Mahoney, active in the 1870's and 1880's, was purchased by the Wildman Company in 1894 and operated with the Wildman until 1906. All three properties passed into control of Lincoln Consolidated Mines in 1906. In 1910 the Wildman and Mahoney were reopened and operated until 1912, since which date they have been idle.

Combined production of the Wildman and Mahoney was slightly less than $5,000,000, while the Lincoln produced $2,200,000 (Bowen and Crippen, 1948, p. 64).

In the area of the three mines the Mariposa slate is interfolded with meta-andesite which has been partially altered to amphibolite schist (Tucker, 1914, p. 37). In the upper levels of the Wildman and Mahoney the ore occurs in a contact vein along the slate-schist contact (Logan, 1934, p. 96). Below the 500-level in the Wildman and the 250-level in the Mahoney the belt of mineralization passes into altered schist. Besides the main belt of mineralization a number of small ore shoots in the schist were developed. The ore bodies pitch southward.

The Lincoln was worked through a 2000-foot inclined shaft, the Wildman through a 1400-foot inclined shaft, and the Mahoney through a 1200-foot inclined and a 616-foot vertical shaft. The three claims were connected by a drift running south from the 2000 level of the Lincoln through the Mahoney to a point under the Wildman shaft. Stoping extended to the 1400 level of the Wildman and to the 1200 of the Mahoney while there was no production below the 500 level of the Lincoln.

Both the Wildman and Mahoney Mines were equipped with 40-stamp mills. (Tucker, 1914, p. 37.)

Moore Mine. Location: sec. 34, T. 6 N., R. 11 E., M.D.M., 1½ miles south of Jackson. Ownership: South Jackson Mining Company, c/o J. Schweitzer, Jackson, California.

This mine was active in the 1880's. During that time it had been opened to a depth of 640 feet. Ore was milled in a 10-stamp mill. From that time it was idle until 1921 when it was reopened by the Moore Mining Company, who operated the mine until 1929. Intermittent prospecting continued until around 1934, since when it has been idle. From 1922 to 1929 production amounted to $564,624. (Logan, 1934, p. 98.)

The Moore vein system occurs in and near the fault contact of a northwest-trending belt of Mariposa slate and greenstone. The dip is to the east. Greenstone forms the hanging wall and slate the footwall. At the deeper levels the vein is badly shattered. The main ore shoot worked during the early period of operation was 1200 feet long and averaged 16 feet in width. It was cut off at a depth of 750 feet by faulting. Small amounts of ore were found on the 1100 level, and between the 1800 and 2100 levels an irregular ore body was mined. Much of the ore found at the deeper levels occurs in badly faulted vein segments and lenses.

Workings consist of a 2291-foot inclined shaft, drifts, crosscuts, and stopes. Many of the underground workings are a result of the extensive prospecting done in the 1920's. During this time several long crosscuts were driven east and west, none of which disclosed any ore. Ore was milled in a 20-stamp mill erected in 1922.

Nevada Wabash Mine. Location: NW ¼ sec. 6, T. 6 N., R. 11 E., M.D.M., adjoining the North Star mine, 1 mile north of Sutter Creek. Ownership: Nevada Wabash Mining Company, Box 504, Sutter Creek, California.

In 1946 the Nevada Wabash Mining Company had a lease on the Wabash mine property. Under the direction of R. V. Kohls a 420-foot southeast adit 600 north of the adjoining North Star mine was cleaned out and extended for a short distance. A 15-inch vein dipping northeast in greenstone was encountered.[9]

There was no production and work was abandoned soon afterward, according to J. Schweitzer, mining engineer.

New Hope Mine. Location: SW ¼ sec. 3, T. 7 N., R. 10 E., M.D.M., 1 mile west of Plymouth. Ownership: Harold and Emma Swingle, Plymouth, California.

By 1890, a shaft 75 feet deep had been sunk on this property. (Irelan, 1890, p 121.) The records show no production from the mine since 1880. In 1947, diamond drilling by the owner indicated an 18-inch-wide ore body averaging 65 dollars per ton in gold values. A new 85-foot vertical shaft was then sunk and a 65-foot drift was run to the south at the 85 level. The small ore body indicated by the drilling consisted of free gold with pyrite in quartz veinlets in greenstone, according to Ernest V. Grant, mining engineer. Additional diamond drilling is under way east of the shaft. The property is intermittently worked by two men.

Newman Mine. Location: NE ¼ sec. 33, T. 7 N., R. 13 E., ¼ mile northeast of the West Point Power House just north of where the highway crosses the Mokelumne River. Ownership: This property was acquired by Pacific Gas and Electric Company, San Francisco, California, for right-of-way for construction of the West Point tunnel.

The outcrops were worked superficially in the early days by the Chinese and Mexicans. This property was located and patented in 1928 by the China Garden Mining Company of Missouri who operated it until 1933. A total of $160,000 in gold had been taken from this mine by the end of 1933 when the mine was closed, according to W. E. Stirnaman, former miner.

The gold occurs in 1-to-10-foot wide quartz veins which strike northwest and are almost vertical. Country rock is granodiorite.

A 1400-foot, north-trending adit has a 200-foot winze 400 feet from the portal. The vein has been stoped out to the surface along the strike 500 feet from the portal. This property was equipped with a 10-stamp mill, concentrators and an air compressor.

° Logan, C. A., unpublished report, 1949.

Oneida Mine. Location: sec. 17, T. 6 N., R. 11 E., M.D.M., 1½ miles north of Jackson. Ownership: South Eureka Mining Company, c/o A. J. Mayman, 400 California Street, San Francisco, California.

One of the important early-day lode mines, the Oneida was producing ore averaging as high as 40 dollars per ton in the 1860's (Logan, 1934, p. 111). It was operated intermittently from that date until the late 1890's when much extensive development work was done and a new mill erected. Operations continued until 1914, since which date it has been idle. Total production of the Oneida was in excess of $2,500,000 (Bowen and Crippen, 1948, p. 64).

The vein occupies a fissure chiefly in slates near the contact with greenstone which lies to the west (Tucker, 1914, p. 39). The main ore shoot is 8 feet wide and 150 feet long and is near the shaft. North of this is an irregular-shaped ore body. A small but rich ore body was developed to the north of this on the 1000 level. Best ore was on the 1500 level, although there was production down to the 2000 level where some coarse gold was found in a thin quartz seam (Storms, 1900, p. 63).

Workings included a 2280-foot vertical shaft in the hanging wall, which cut the vein at the 1900 level, a 1350-foot inclined shaft, and two other shallow inclined shafts. The Oneida was connected with the South Eureka mine on the 1800-foot level (Tucker, 1914, p. 39).

Ore was milled in a 60-stamp mill. When in full operation 40 or 50 men were employed at the mine and mill.

Original Amador Mine. Location: sec. 36, T. 7 N., R. 10 E., ¼ mile north of Amador City. Ownership: J. W. Bullock, c/o C. E. Crandall, North Bend, Washington.

This mine, a consolidation of six claims, was originally opened in 1852. By 1872 workings extended to a depth of 365 feet and a 40-stamp mill was erected. From 1874 to 1898 little work was done other than intermittent prospecting. Between 1898 and 1918 the Original Amador Consolidated Mines Company operated the mine. From 1910 to 1918 production was between $90,000 and $130,000 per year (Logan, 1934, p. 104).

The property lay idle until early in 1935 when it was reopened under the direction of Hamilton, Beauchamp, and Woodworth. The mill was reconditioned and its capacity was increased. Operations were suspended in the middle of 1937, since which date the mine has been idle.

After the mine closed the mill was operated until 1942 by Keystone Mines Syndicate to treat ore mined at the Keystone, according to T. S. O'Brien. Estimated total production for the Original Amador is $3,500,000 (Bowen and Crippen, 1948, p. 65).

Near the surface, the main Original Amador quartz vein ranges from 20 to 50 feet in thickness (Tucker, 1914, p. 39). The vein is represented by two branches which are separated by greenstone. The footwall of the west vein is thinner and of better grade than the hanging-wall branch. The thickness of this vein ranges from a thin fissure to 5 feet of solid quartz (Knopf, 1929, p. 57). The maximum width of the hanging-wall branch is 90 feet. The hanging wall is greenstone while the footwall is slate. Below the 300 level the veins enter the greenstone and assume a flatter dip. Ore was obtained in part from the Original Amador quartz vein and in part from large irregular masses of auriferous greenstone occurring in the hanging-wall of the quartz vein (Knopf, 1929, p. 57).

Workings consisted of a 1238-foot inclined shaft and over nine miles of drifts, crosscuts, and raises. Because of the firm wall and width of ore, shrinkage stoping was used on the main vein (Logan, 1934, p. 105).

The mill at the Original Amador mine had a capacity of 300 tons per day. Ore was crushed, screened, run through a 20-stamp mill, and concentrated on Diester tables. Concentrate was then reground in Hardinge ball mills and cyanided.

Peterson Mine. Location: NE ¼ sec. 5, T. 6 N., R. 12 E., about one mile southwest of Pine Grove. Ownership: William F. Peterson, Pine Grove, California.

In 1931, W. F. Peterson purchased the mine from J. Cuneo and operated it continuously until the government wartime restrictions closed the mine in 1942. Total gold output between the years of 1931 and 1942 amounted to approximately $250,000, according to Mr. Peterson. It was not until 1948 that production was once again resumed, this time under the guidance of H. P: Livingston, president of the Lomar Milling Company.

The ore consists of leaf gold and disseminated gold in quartz with associated pyrite and arsenopyrite, and occurs in the contact zone between greenstone and granite. Ore deposition appears to be controlled by block faults in which the dominant strikes are northeast. The largest vein and secondary veins have a north strike while other important veins strike east. All of the veins are nearly vertical. One highly mineralized shear zone about 50 to 60 feet wide was encountered 560 feet from the portal of the 200 level. Selected ore assays $40 to $60 and better per ton while the mill run ore assays about $30 per ton, according to Mr. Livingston.

Most of the present development work and the small amount of stoping has been confined to an adit which is termed the 65-foot level. There are two other caved adits which are termed levels, the 200 and 270. The ore is trammed in a small ore car by hand out of the 65-foot level and deposited in a storage bin before it is hauled to the mill by truck. Reopening work started in 1952 on the old 200-foot level which constituted the original adit.

About a quarter of a mile southeast of the mine, the company maintains an assay laboratory and a mill. Ore is run through a jaw crusher and then a 5-stamp mill. Fines are directed over an amalgamation plate while the tailings are impounded.

Since reopening of this mine in 1948, the mining operations have concentrated on development work, consequently, only a small production of gold has been recorded. H. P. Livingston and W. F. Peterson do all the mining and milling.

Pine Grove Mine. Location: SW ¼ sec. 23, T. 7 N., R. 13 E., M.D.M., one mile southeast of Buckhorn Lodge. Ownership: Wendell M. Miller, et al., c/o E. A. DeRuchia, Reno, Nevada.

This mine was originally developed by Wendell M. Miller and associates about 1935 and was worked intermittently until 1941. From 1944 to January 20, 1950 the mine was operated by Associated Metals, Inc. During this time 40 to 50 tons of ore assaying up to 80 dollars per ton were produced, according to Sheriff Harry James, former operator. The mine has been inactive since January, 1950.

Free gold associated with pyrite, chalcopyrite, and galena occurs in a 1- to 5-foot quartz vein in granodiorite. The vein strikes northwest and dips steeply west. Some diamond drilling was done in 1946. Hayes Evans reported that during the last months of operation the ore assayed $13.00 in gold and $8.00 in silver.

Workings consist of a 90-foot inclined shaft, several hundred feet of drifts, and 100 feet of crosscuts. There are two levels, the 45-foot and 90-foot levels. An old shaft 400 feet to the south was connected to the main workings, according to Mr. James.

Pine Grove Unit Mine. Location: SE ¼ sec. 32, T. 7 N., R. 12 E., M.D.M., ¼ mile southwest of Pine Grove and adjoining the Peterson mine. Ownership: W. F. Peterson, Pine Grove, California.

Work started on this property in 1938 by W. F. Peterson. H. P. Livingston leased the property in 1946 and operated the mine until 1948. Mr. Livingston reported that in the later 1930's, $5,000 in gold was produced.

The ore consists of free gold associated with smaller amounts of bornite and sphalerite in a quartz vein which strikes east and dips about 55° south. The vein varies from a few inches to 3 feet in width. Country rock is Calaveras slate. Eight hundred tons of $27.00 per ton gold ore is blocked out at the 60-foot level, according to Mr. Livingston.

A 187-foot inclined shaft was sunk on the vein. Other development work consists of 200 feet of east-trending drifts on the 60-foot level and several hundred feet of drifts on the 120-foot level.

Pioneer Lucky Strike Mine. Location: SE ¼ sec. 20, T. 7 N., R. 13 E., M.D.M., 1¾ miles northeast of Pioneer Station. Ownership: J. H. Hauhuth, Pioneer Station, California.

Discovery work on this property began in 1921 by Rudolph Moar and brothers. After the late 1920's the mine was worked almost continuously by several different lessors until 1941. The mine produced $300,000 in gold during the period of 1921 to 1941, according to Mr. Moar.

Free gold is associated with pyrite in three ore shoots. The shoots are located in a quartz vein which varies in width from 6 inches to 5 feet, the average being 28 inches. The veins strike N. 50°-55° W. and dip steeply to the southwest. Country rock is granodiorite.

Workings consist of an 1100-foot southeast adit and an 180-foot winze sunk 170 feet in from the portal. Most of the ore from the adit to the surface has been stoped.

Ore was run through a 30-ton mill. The owners reported that 30 men were employed at the mine and mill when in full operation.

Plymouth Consolidated Mine. Location: sec. 11, T. 7 N., R. 10 E., M.D.M., in the city limits of Plymouth. Ownership: B. Monte Verda and E. C. Taylor, 369 Pine Street, San Francisco, California.

The claims that formed the Plymouth mine were originally located in 1852. They were worked separately until consolidation in 1883 under the name of Plymouth Consolidated Mining Company.

In 1888 a severe fire was not brought under control until the mine was flooded. The mine was idle from 1892 to 1911. In 1911 the Plymouth Consolidated Mine Company, Limited, was incorporated and acquired the property. In 1925 it was sold to the Argonaut Mining Company (Logan, 1934, p. 107).

FIGURE 12.   Plymouth Consolidated mine, Empire shaft.

Little work was done between 1928 and 1939 but a combined flotation and cyanide plant was put into operation by the Argonaut Mining Company to treat tailings. Frank Standridge reported that during this time, a small amount of exploration was done in the upper levels, principally the 400 and 500 levels, of the Empire shaft. This operation shut down in 1943.

In 1946, the Empire workings of the Plymouth Consolidated mine were reopened by the Argonaut Mining Company. This was a resumption of prospecting interrupted by World War II. On the 400 level an old drift was extended north for about 700 feet. Only a small amount of pocket ore was encountered and work ceased early in 1947, according to Mr. Standridge.

The Plymouth is the most northern of the major mines of the Mother Lode in Amador County. Total production is estimated to be in excess of $13,500,000 (Bowen and Crippen, 1948, p. 65).

There are three parallel vein systems called the Empire, Reese, and Woolford and several hanging-wall contact veins in a mile-wide belt of Mariposa slate; the veins have a general north strike and dip 65° east (Tucker, 1914, p. 41). The Empire vein system contains the main ore shoot which varies from 4 to 16 feet in width. In the deeper levels ore is composed chiefly of a ribbon structure of slate and quartz.

Prior to 1923, the main or Pacific shaft was sunk to a depth of 3400 feet. At that time ore had been mined through a winze, hoisted to the 3400 level, trammed to the Pacific shaft, and then hoisted to the surface. In 1925, the Pacific shaft was sunk to a depth 4450 feet, of which the first 1600 feet is vertical and the balance on an incline of 60 degrees. The principal ore body between the winze and shaft has been stoped out as far down as the 4300 level. Some of the last work done before the mine closed in 1928 was between the 4450 and 4300 level (Logan, 1934, p. 107).

The Empire ore shoots were worked through the 1280-foot inclined Empire shaft. During the last phases of operations this was reconditioned to the 500 level. Approximately 350 feet north of this shaft is the 1065-foot North Empire shaft.

The Plymouth mine was the first in the Mother Lode to have ore delivered to the mill from the shaft by conveyor belt (Logan, 1934, p. 195). Ore was then delivered to 30 stamps and then to cone separators to separate the fines, while the coarse feed was reground in two Hardinge ball mills. The pulp was then sent to amalgamation plates, classifiers, Wilfley tables, and vanners. Capacity was 420 tons per day.

Rainbow Mine. Location: NE ¼ sec. 32, T. 7 N., R. 12 E., M.D.M., 1½ miles northwest of Pine Grove. Ownership: Claude Hanley, 1727 Hiawatha Avenue, Stockton, California.

Prior to 1935, this property was a small prospect. In 1935, C. Evans and W. W. Evans of Glendale purchased the property and began to sink a shaft. The mine was worked continuously from 1935 until the fall of 1941 but has since been idle.

Ore consists of free gold in quartz with pyrite and galena. The quartz vein which varies from 8 to 14 feet in width, strikes east and dips to the north. Several north-striking quartz stringers branch off the main vein. Pocket gold is very characteristic at this mine. Country rock is Calaveras slate.

In 1935, an inclined shaft was sunk to a depth of 157 feet where it intersected the main ore shoot. This shaft caved and a new inclined shaft 150 feet east was sunk to a depth of 200 feet. At the 200-foot level off the new shaft, a drift was driven about 200 feet to the west, according to Mr. Ruffino. There are a number of buildings on the property but all the equipment has been removed.

Red Hill Mine. Location: NE ¼ sec. 32, T. 7 N., R. 12 E., M.D.M., 1½ miles northwest of Pine Grove. Ownership: Claude Hanley, 1727 Hiawatha Avenue, Stockton, California.

A gold-quartz vein was discovered on this property in 1946. Development work continued intermittently until 1949 since which date the property has been idle. Prior to 1949 the mine was under lease to H. P. Livingston of Pine Grove and approximately $5000 in gold was produced, according to Lee Gardner.

Ore consists of free gold in quartz associated with pyrite and small amounts of chalcopyrite and bornite. The nearly vertical gold-quartz veins strike north and occur in sheared and fractured zones within Calaveras slates and schists.

An 84° inclined shaft was sunk 120 feet. At the 100-foot depth a 58-foot drift was driven north. At the 50-foot depth two crosscuts were driven, one 30 feet east and the other 100 feet west. A 20-ton mill, which was erected on the Red Hill property but never operated, was moved to the Italian mine near Drytown. Four to six men were employed at the mine, according to Mr. Gardner.

South Eureka Mine. Location: sec. 17, T. 6 N., R. 11 E., M.D.M., one mile south of Sutter Creek. Ownership: South Eureka Mining Company, c/o A. J. Mayman, 400 California Street, San Francisco, California.

This mine was originally developed in 1891 and during much of its period of operation was one of the largest producers in the country (Logan, 1934, p. 112). Until 1908 production was moderate from small ore bodies in the hanging-wall vein. In 1908 the footwall vein system was discovered. From this date until the mine closed in 1917, over 980,000 tons of ore were produced (Logan, 1934, p. 113).

In 1921 the Central Eureka Mining Company took an option on the property and did some exploratory work from the Central Eureka mine into the South Eureka mine (Logan, 1934, p. 112). Also for a time they kept part of the workings unwatered and the shaft in repair. Total production of the South Eureka was $5,300,000 (Bowen and Crippen, 1948, p. 64) from which about $1,000,000 in dividends were paid.

At this mine the Mother Lode system occurs in and along a belt of Mariposa slate which is several hundred feet wide. Greenstone (meta-andesite of the Logtown Ridge formation) is on both sides of this slate belt (Knopf. 1929, p. 63). Three veins were developed: the hanging-wall. footwall, and middle veins (Tucker, 1916, p. 47). The hanging-wall vein is on the slate-greenstone contact to a depth of 2000 feet where the vein passes into the greenstone. Ore occurs in bunches in slaty gouge, in pockets, streaks in greenstone, and at the junction of the foot-wall vein and the middle vein below the 2000-foot level. The chief mine production was from this ore body between the 2750 and 2000 foot levels.

The mine was developed through a 2785-foot inclined shaft. Cross-cuts were run east and west to the ore bodies. A winze was sunk to the 2900-foot level but no ore was found. When the Central Eureka had the property under option they extended several of the deeper levels of the Central Eureka mine into the South Eureka, the deepest being the 4100-foot level. The result of this work was disappointing as no ore was found (Logan, 1934, p. 112).

Ore was milled in an 80-stamp mill with a capacity of 400 tons per day. The ore was crushed. run through a 24-mesh screen, amalgamated, and concentrated on 48 Frue vanners. Mill recovery ranged from 85 to 91 percent. At full production, 200 to 250 men were employed at the mine and mill.

South Spring Hill Mine.   Location: sec. 31, T. 7 N., R. 11 E., M.D.M., ½ mile south of Amador City. Ownership: Keystone Mining Company, c/o John D. Culbert, Amador City, California.

This mine was located in 1851 and worked continuously until 1893. In 1887, a 30-ton stamp mill was crushing 70 tons of ore daily (Logan, 1934, p. 91). The mine was reopened in 1900 and closed again in 1902. In 1920, title to this property was acquired by the Keystone Mining Company. Unwatering of the South Spring Hill shaft began in the middle of 1934 by the Keystone Mines Syndicate in conjunction with the reopening of the Keystone mine. The South Spring Hill and Keystone shafts were retimbered and connected by underground workings. Since the company was primarily interested in mining ore from the Keystone, no ore was produced from the South Spring Hill mine during this time. Although there was work done at this mine from 1934 to 1942, there has been no recorded production of ore since 1902. Total production from 1888 to 1902, inclusive, amounted to $1,092,472 (Logan, 1934, p. 91).

The gold-quartz vein strikes northwest, dips 52° to 85° northeast, and is about 20 feet wide; its greatest width measures 50 feet. Country rock consists of a slate footwall and a greenstone hanging-wall. A heavy gouge on the footwall is a common occurrence.

The property has been developed by three inclined shafts, South Spring Hill, Talisman, and Medean, which are 1200, 600, and 600 feet deep, respectively. Talisman and South Spring Hill shafts are about 2300 feet apart and connected underground. These shafts were sunk on the contact vein while the Medean shaft was sunk on a parallel lead to the east in greenstone.

Treasure (Hazard) Mine. Location: sec. 25, T. 7 N., R. 10 E., M.D.M., one mile north of Amador City. Ownership: Treasure Mining Company, 220 Montgomery Street, San Francisco, California.

Prior to 1867 the Treasure mine had produced 5000 tons of ore. It lay idle until 1907 when the Treasure Mining Company reopened it. It was operated until 1922, since which date it has been idle. Total production of the Treasure mine was about $1,000,000; however, no dividends were paid as the surplus was exhausted looking for more ore (Logan, 1927, p. 184).

The Treasure vein, in which there are two ore shoots, is in slate to the 400 level where the vein passes into schist. The dip which is to the east, is shallow near the surface and steepens to 65° to 70° in the schist. The two shoots average six feet in width and vary from 90 to 400 feet in length (Tucker, 1914, p. 50). A disseminated ore body 700 feet long and 18 feet wide in schist was developed on the 1000 level. The best ore was above the 2320-foot level below which a fault cut off most of the ore.

Workings consisted of a 1600-foot inclined shaft, and winzes which reach an inclined depth of 3030 feet. Levels were 160 feet apart. The 1600 level was connected with the 1500 level of the Bunker Hill to the south (Logan, 1934, p. 182).

Treatment of the disseminated ore differed from the standard practice on the Mother Lode. Ore was run over a one-inch grizzly with the oversize being sent to a gyratory crusher and then to a trommel. The oversize from there was sent to a jaw crusher. Coarse ore was conveyed to a large Hardinge mill while fine ore was conveyed to a smaller one. From the ball mills it passed to a screen and shaking plates. Pulp from the plates was sent to classifiers, Wilfley tables, concentrators, and Frue vanners. Capacity was 150 tons in 24 hours.

Valparaiso (Black Metal) Mine. Location: sec. 10, T. 5 N., R. 11 E., M.D.M., 4½ miles south of Jackson. Ownership: Valparaiso Mining Company, Jackson, California.

This mine was worked prior to 1888 and gold ore was produced until 1898. From 1898 until 1931, the mine is credited with producing $100,-000 in gold (Logan, 1934, p. 114). Gene Boro reported mining gold worth $600 in 1937. From 1937 to 1939, a partnership composed of Leo Dunlavy, Charles Spencer, and Dean Wiley mined the property intermittently. Approximately $5000 in gold was produced from small pockets from 1937 to 1939, according to Mr. Dunlavy. The mine has been idle since 1939.

Gold occurs in pockets, quartz veins, and quartz stringers in greenstone and Mariposa slate. The workings are concentrated in a heavily faulted area along the slate-greenstone contact, and consist of a 1300-foot adit, drifts, crosscuts, raises, winzes, and stopes, according to Mr. Boro.

Wonder Mine. Location: sec. 36, T. 7 N., R. 10 E., M.D.M., about 150 feet east of the Keystone mine shaft. Ownership: Keystone Mining Company, c/o John D. Culbert, Amador City, California.

This property was leased from the Keystone Mining Company by a partnership consisting of Martin Ares, George Thurman, and John Rodriquez in January of 1952. At that time, the operators began sinking an inclined shaft which they called the Wonder mine. Mining operations were suspended in May of 1952 when the new workings ran into the old workings of the Keystone mine at a depth of 65 feet.

The ore was a weathered, iron-stained quartz containing pyrite, and exceeded $25 per ton in gold, according to Mr. Ares.

Zeila Mine. Location: sec. 28, T. 6 N., R. 11 E., M.D.M., on the southeast side of Jackson. Ownership: Mark and Frances Eudey, Martell, California.

The Zeila produced over $5,000,000 in gold. It was first worked in the 1860's and by 1867, there was a 16-stamp mill crushing the ore and a chlorination plant treating the concentrates. The mine was closed in 1875 but reopened again in 1880 and operated continuously until 1914 (Logan, 1934, p. 115). The mine has been inactive since 1914 and subsequent to that date it passed into possession of the Kennedy Mining and Milling Company. Mine tailings are utilized by the Amador County road department for use as fill from time to time.

Ore occurs in quartz stringers on the contact of greenstone and Calaveras slate and schist. The vein, which averages 20 feet in width, dips 65° east (Tucker, 1914, p. 52). Gouge occurs on the footwall of the main vein only and ranges from a few feet to fifty feet in thickness. The highest grade ore occurs in the quartz-ribbon structure.

The shaft was sunk 1700 feet on an angle of 65°. On the 1570-foot level, a drift was driven 3000 feet north from the shaft, at which point a winze was sunk on the vein to a depth of 458 feet. One hundred and fifty-seven feet below the 1570 level, a 352-foot drift was driven north on the vein from the winze. Two hundred and ninety-five feet below the 1570 level, a 450-foot drift was driven north.

Gold Mine Tailings

Argonaut Tailings. Location: sec. 20, T. 6 N., R. 11 E., M.D.M., one mile north of Jackson. Ownership: B. Monte Verda and E. C. Taylor, Box 367, Jackson, California.

From 1923 until October, 1938 the Amador Metals Reduction Company under the direction of Hamilton, Beauchamp, and Woodworth, consulting engineers of San Francisco, treated the tailings of the Argonaut mine for the gold content. During this operation about 70,000 tons of tailings per year were treated with a recovery averaging $60,000 per year, according to S. E. Woodworth.

Tailings were hydraulicked from the tailings pile into a classifier. Both sands and slimes from the classifier were pre-coated with coal tar to prevent pre-precipitation of the dissolved gold in the cyanide solution because of the carbon present. The sand was then treated in leaching vats while the slimes were directed to an agitator and filtered on Oliver filters. Sand and slime residues were discarded. The gold-bearing solution was then precipitated, roasted, and the gold melted into bullion. When in operation, six men were employed at the plant.

Central Eureka Tailings. Location: SE ¼ sec. 7, T. 6 N., R. 11 E., M.D.M., at the Central Eureka mine in Sutter Creek. Ownership: Central Eureka Mining Company, 2210 Russ Building, San Francisco, California.

In 1930, a partnership consisting of Hamilton, Beauchamp, and Woodworth, consulting engineers of San Francisco, and Glen O'Brien of Ione took an option on the Central Eureka mine tailings. The plant operation started in 1932 and continued steadily until 1938. Eight hundred and eighty thousand tons of tailings were treated, according to Mr. O'Brien. Assay values varied from $1.50 to $3.00 per ton, on the basis of $35 per ounce gold. Recovery approximated 70 percent.

The tailings pile was sluiced by hydraulic means to a classifier through a sand pump. Both sands and slimes were then treated with three pounds of coal tar per ton to prevent pre-precipitation because of the carbonaceous content. The sand was treated in leaching vats on an eight day cycle. The slimes were directed to an agitator, thickener and filters. Both sand and slime residues were discarded. Precipitation was achieved by treating the solution with the Merrill-Crowe process. Precipitates were then acid treated, roasted, and melted into gold bars. Mr. O'Brien reported that capacity of the plant was 600 tons per day and that it was operated by 15 to 20 men.

Delta Placer Gold Company. Location: N ½ sec. 10, T. 5 N., R. 9 E., M.D.M. (projected), on the Arroyo Seco land grant, for about 2 miles along Jackson Creek. Ownership: Glen O'Brien, Ione, California.

Tailings from the mines around Jackson, including the Zeila, Argonaut, and others, were diverted into Jackson Creek before enactment of the anti-debris or Caminetti Law of 1893. During the winter time, the high waters of Jackson Creek transported and redeposited the tailings in Jackson Valley to an average depth of 2½ feet, although in some places, the depth was 8 feet.

The Delta Placer Gold Company began operations at the present location in 1935 and continued without interruption until 1942. One million three hundred and fifty thousand tons of redeposited tailings were treated with a recovery exceeding 80 percent, the owner reported. Gold assays varied from 80 cents to $2.00 per cubic yard in gold values. The property has been idle since 1942.

The material was bulldozed and conveyed to a pulper and wet-screened with a Symons vibrating 10-mesh screen. The oversize was discarded while the minus 10-mesh material was pumped a maximum distance of two miles to be classified into sands and slimes. Leaching of the sands was accomplished by cyanidation while the slimes were directed to a Dorr agitator where gravity separation of the gold was effected. The slimes were run through a counter-current decantation process and the residues discarded. Eighteen men were employed by the company.

During the stripping of the re-deposited tailings, approximately 220 acres of farming land were reclaimed along Jackson Creek. Discarded material was used to construct a levee on Jackson Creek.

FIGURE 13.   Delta Tailings Company, 1936. *Photo by Glen O'Brien.*

Kennedy Tailings.   Location: SW ¼ SW ¼ sec. 16, T. 6 N., R. 11 E., M.D.M., at the Kennedy mine north of Jackson. Ownership: Mark and Frances Eudey, Martell, California.

In 1948 and 1949, for a period of 1½ years, the Kennedy mine tailings were worked intermittently by Frank Fuller, Jr., of Jackson, who reported that approximately 20,000 tons of gold-bearing tailings were treated with an average recovery of $2.00 per ton.

Fines from the trommel were jigged, run over vibrators, jigged again, and passed over Wilfley tables to recover the fine gold. One to three men operated the washing plant.

Plymouth Tailings.   Location: sec. 11, T. 7 N., R. 10 E., M.D.M., at the site of the Plymouth Consolidated mine. Ownership: B. Monte Verda and E. C. Taylor, Box 367, Jackson, California.

During the period from 1939 to 1943, the Argonaut Mining Company, which then owned title to the Plymouth Consolidated mine, operated a flotation and cyanide plant to process Plymouth mine tailings. This plant had a capacity of 500 tons per 24 hours. Tailings were run through a sand pump, screened, flumed to agitators, and then to flotation cells. Concentrates were then shipped to the smelter. Flotation tailings were classified, cyanided by leaching, and gold was precipitated in tanks. About 750,000 tons of tailings were treated in this plant.[10]

Placer Gold

The gold of Amador County placers originated in quartz veins deposited deep in the Bedrock series of the Sierra Nevada. The gold was liberated from these veins by prolonged erosion of the roof rocks and by disintegration of the wall rocks, chiefly under unusual conditions of chemical decay. Gold, one of the relatively indestructible minerals, be-

[10] Logan, C. A., unpublished report, 1948.

came intermingled with other weather-resistant material in streams and was concentrated by action of running water. Owing to its great weight gold tends to accumulate at or near the base of such deposits.

Although there are river gravels of many different ages in the Sierra Nevada, not all of them contain gold in paying quantities. The conditions under which the economic deposits formed are discussed in another part of this report under Geologic History. Several distinct types of river gravel deposits can be recognized:

1. Eocene ("Tertiary") gold-bearing gravels typified by an abundance of white quartz and a paucity of volcanic debris (excepting old metavolcanics). These are almost always found in remnants of channels worn directly into crystalline bedrock or on terraces adjacent to such channels. They commonly are overlain by white rhyolite ash.

2. Inter-rhyolitic channel gravels (Miocene) containing abundant light-colored rhyolitic debris as well as white quartz, but no dark-colored andesitic debris. These commonly are found in channels cut into Eocene gravels or even in channels cut into rhyolite ash. These tend to be much less productive of gold than the Eocene gravels although they occasionally robbed rich Eocene channels and themselves became locally productive.

3. Inter-andesitic channel gravels (Mio-Pliocene) containing abundant andesite debris as well as debris of all older rocks. The chances of these gravels being rich in gold are slight. They may occur by themselves or in super-position above inter-rhyolitic and Eocene deposits.

4. Recent stream gravels consisting predominantly of abrasion-resistant rocks of all types. These may be rich in gold in streams that drain areas where quartz veins are being actively eroded or in areas favorable for re-concentration of materials from older gravels. They are always found along or close to present-day streams.

Associated with gold in placer deposits are heavy mineral grains which are also resistant to chemical and mechanical destruction. These grains, commonly grouped under the heading of black sands, consist essentially of magnetite with varying amounts of ilmenite, rutile, platinum group metals, zircon, chromite, garnet, and other heavy, weather-resistant minerals.

Garibaldi Dredge. Location: SE $\frac{1}{4}$ sec. 23, T. 7 N., R. 12 E., M.D.M., $\frac{1}{2}$ mile east of Volcano. Ownership: Frank and Peter Garibaldi of Volcano.

This property, located at the junction of Pioneer and Sutter creeks, has been worked intermittently since 1940 by the owners. Mr. Frank Garibaldi reports that the stream gravels range from 5 to 15 feet in thickness and rest on a bedrock of Calaveras limestone.

In the process of working an area of ground, the topsoil is first bulldozed to one side. The gravel is then delivered by dragline to a track-mounted washing plant which has a capacity of 40 cubic yards per hour. It is equipped with a 28-foot trommel and a 50-foot stacking boom. Gold recovery varies from 10 to 20 cents per yard. After an area has been worked the gravel piles are leveled and the topsoil is then bulldozed over the gravel.

Lilly Dredge. Location: sec. 24, T. 8 N., R. 9 E., and secs. 20, 28, and 29, T. 8 N., R. 10 E., M.D.M., $5\frac{1}{2}$ miles northwest of Plymouth on the Cosumnes River. Ownership: dredge owned by E. L. Lilly, 1640 East Poplar Street, Stockton 5, California.

From September, 1938, to August, 1942, and again from July, 1946, to November, 1949, gold-bearing gravels on the Cosumnes River were worked by the Lilly dredge. The pay gravel has been worked for a distance of $3\frac{1}{2}$ miles along the river from the southeast corner of sec. 25, T. 8 N., R. 9 E., to the northwest corner of sec. 28, T. 8 N., R. 10 E., M.D.M. Approximately 2,125,000 cubic yards of gravel averaging 18 cents in gold per yard were handled, according to Mr. Lilly.

Gravel was delivered by a model 85 Northwest dragline having a $2\frac{1}{2}$-yard bucket to a Bodinson floating washing plant. The washing plant, powered by a diesel electric generator, was equipped with 52" x 32' trommel, Hungarian riffles set in sluices, and a 70-foot stacking belt. Gold was melted into bars in the cleanup house and sent to the mint.

Madrill Dredge.   Location: sec. 1, T. 4 N., R. 9 E., M.D.M., on the north side of the Mokelumne River 5 miles southwest of Buena Vista. Ownership: C. R. and Annabelle C. Brown, R. F. D., Ione, California.

In the spring of 1952 operations were begun by George Madrill to work terrace gravels along the Mokelumne River with a floating dragline dredge. The gravels average 20 feet in thickness and are about 65 feet above river level.

Gravel is delivered to the washing plant by a dragline suspended from a 100-foot boom mounted on the dredge. The dredge is equipped with a 72-inch trommel, a 50-foot stacking boom, and a 460-horsepower diesel-electric power plant. Capacity is 150 cubic yards per hour, according to George Madrill.

Pigeon Drift Mine.   Location: SW $\frac{1}{4}$ sec. 28, T. 8 N., R. 11 E., M.D.M., one mile northwest of Fiddletown. Ownership: Fred N. Pigeon, Fiddletown, California.

Work was first started on the Pigeon drift mine during the 1850's. Intermittent operations by the Pigeon family continued until 1935. At that time the property was leased to F. W. Nowell who operated it until 1942. It has been idle since then except for some re-timbering done by Harry Gould of Sacramento after World War II. From 1935 to 1942, $18,000 in gold was produced, according to Mr. Pigeon.

This deposit consists of a northeast-striking Tertiary stream channel of cemented sands and gravels capped by Tertiary andesites. The channel which lies on a bedrock of Calaveras slates varies from 20 to 50 feet in width and is about 6 to 8 feet in depth. Assays average about $3.00 per cubic yard.

The workings consist of a 300-foot southeast-trending adit driven into the channel and several hundred feet of drifts within the channel. The cemented gravel was crushed, run through a trommel, washing plant, and a sluice box.

Secor Dredge.   Location: S$\frac{1}{2}$ sec. 18, T. 7 N., R. 12 E., M.D.M., four and one-half miles southeast of Fiddletown. Ownership: Elmer Evans, c/o Treasurer's Office, Jackson, California.

Mr. H. R. Secor of Los Angeles has been intermittently working the property on a royalty basis since the summer of 1951. Placer gold is recovered by means of a dryland dredge.

This deposit consists of well-cemented Tertiary auriferous gravel which ranges from a few inches to four feet in thickness, overlying a greenstone bedrock.

A $\frac{1}{2}$-yard dragline delivers the gravel into a 3- by 6-foot trommel. The undersize falls into a mechanical double rocker, over riffles and into

FIGURE 14.   Secor placer near Fiddletown. *Photo by Mary Rae Hill.*

a quicksilver sluice. Capacity of this washing plant is 100 yards per day, according to Mr. Secor.

Iron

Although there has been no recorded production of iron ore from Amador County, deposits of low-grade iron occur as irregular patches and cappings in the western part of the county, particularly in the Ione district. Some work was done on the Thomas iron deposit in mid-1952 but no other deposit in the county has been developed.

The largest known deposit occurs on Rancho Arroyo Seco 3 miles southwest of Ione in secs. 27 and 28, T. 6 N., R. 9 E., M.D.M. (projected). Limonite and some hematite act as cementing agents in sandstone of the Ione formation. A number of patches in both sections occur as cappings on small knolls. The largest outcrop has a width of about 300 feet from east to west, and a length of about 1100 feet north to south. The thickness indicated by several exposures where a little digging has been done is only about 2 feet. A general sample gave on analysis: 34.82 percent iron, 0.211 percent phosphorous, 18.5 percent silica, 0.05 percent manganese, and 0.066 percent sulfur.[11]

In the Clinton peak area, SE ¼ sec. 8 and SW ¼ sec. 9, T. 6 N., R. 12 E., M.D.M., 2 miles southwest of Pine Grove is another iron deposit. This material also occurs as a capping and consists of dark brown limonite with some hematite. The average of two samples taken from a road cut was 27.78 percent iron, 0.150 percent phosphorous, 39.88 percent silica, 0.05 percent sulfur, and 0.90 percent manganese.[12]

11 Logan, C. A., unpublished report, 1949.
12 Logan, C. A., unpublished report, 1949.

Thomas Iron Deposit (Anderson Lease). Location: NW ¼ of SW ¼ sec. 16, and in NE ¼ and SE ¼ sec. 17, T. 5 N., R. 10 E., M.D.M. This property comprises a total of 197 acres north of the Buena Vista-Pardee Dam Road, a mile southeast of Buena Vista. Ownership: George M. Thomas of Jackson, owner of the property leased it to William Anderson on May 13, 1952.

The deposit consists of flat-lying beds of limonite, and lateritic sands and clays of the Ione formation. Several holes drilled to a depth of 18 feet by the H. Earl Parker Company of Marysville, revealed a limonite capping ranging from 2 to 15 feet in thickness.

William Anderson contracted with the H. Earl Parker Company of Marysville to conduct preliminary mining operations. The company worked about 10 days drilling, bulldozing, and clearing brush, and then removed all of their equipment from the property. The original plans were to ship the iron ore to Japan, according to the owner. The property is now idle.

Manganese

The war years 1914-1918 initiated the exploitation of manganese deposits in Amador County. This exploitation ceased in 1919 and was revived by the stimulating effects of World War II. During World War II, the federal Metals Reserve Company bought and stockpiled manganese ore. Between 1919 and 1942 and since 1944, there has been no recorded production of manganese from this county.

For additional information regarding manganese in Amador County, the following references are suggested:

Bradley, W. W. Huguenin, E., et al., 1918, Manganese and chromium in California: California State Min. Bur. Bull. 76, pp. 29-30.

Jenkins, O. P., 1943, Manganese in California: California Div. Mines Bull. 125, pp. 75, 102-103.

Trask, Parker D., 1950, Geologic description of the manganese deposits of California: California Div. Mines Bull. 152, pp. 27-35.

Manganese in Amador County was originally deposited with silica as a chemical sediment, either as a manganiferous silica gel or as a carbonate if sufficient carbon dioxide was present. The manganese is associated with the rocks of the Calaveras (Paleozoic) and Amador (Jurassic) groups. Taliaferro (1943, p. 331) states that the source of both the manganese and the silica was submarine springs having a volcanic source rather than from the leaching of volcanic rocks. Rhodochrosite and bementite were the original ore minerals deposited. Subsequent metamorphism altered some of the original material into rhodonite while continued weathering oxidized the earlier formed manganese minerals into black manganese oxides.

Manganese is distributed through the western portion of Amador County. The manganese ore consist of pods, lenses, or disseminations of oxides and rhodonite in recrystallized chert of the Calaveras series. Most of the deposits are east of the Mother Lode. West of the Mother Lode, a few deposits occur in metachert that is associated with the metavolcanic rocks of the Amador group. Other deposits are concentrations of low-grade manganese oxides in lateritic soils beneath Tertiary lava caps, while still others consist of beds of manganiferous and ferruginous chert in slate.

Lubanko Prospect. Location: SE ¼ sec. 10, T. 7 N., R. 11 E., M.D.M., about 3 miles southeast of Fiddletown. Ownership: Louis Lubanko, Fiddletown, California.

Dr. J. T. Stacy of Pine Grove leased this property during World War II. A few truck loads of manganese ore were mined and shipped as part of a shipment from the Stirnaman mine.

Lenses of psilomelane and pyrolusite, one to three feet thick, occur in massive metachert. These lenses grade into low-grade ore. An ore body 80 feet in length has been exposed by development work. According to Trask, (1950, p. 31) the small amount of ore which was shipped averaged 52 percent manganese, 3 percent iron, and 4.5 percent silica.

The development work consists of two fairly large cuts and two small adits. This property has been idle since World War II.

Perini (Peyton et al. Lease) Mine. Location: NW ¼ sec. 35, T. 7 N., R. 12 E., M.D.M., 2 miles east of Pine Grove. Ownership: Benjamino Perini, Pine Grove, California.

Less than one carload of manganese ore was mined and shipped during World War I (Trask, 1950, p. 32). During World War II, the property was leased to the Garibaldi Brothers and Frank Simpson of Volcano who mined and shipped several carloads of ore of unknown grade.

One large cut on the sidehill adjacent to the shaft exposed pockets of psilomelane with some pyrolusite and rhodonite enclosed in an 8-foot wide band of manganiferous metachert of the Calaveras formation. This same cut exposes a zone of 10 feet containing manganese minerals; the zone has a steep dip to the east and a general northerly strike.

The ore now exposed would probably average about 20 percent manganese, but pockets of ore containing 40 percent could probably be found (Trask, 1950, p. 32). A 25-foot gallows frame stands over a shaft 74 feet deep. On the 40-foot level a 25-foot crosscut has been driven to the east. There are extensive bulldozing cuts on this property. A quarter of a mile south of this main shaft is another small shaft.

Peyton (Crocker-Preston) Prospect. Location: SW ¼ sec. 15, T. 7 N., R. 12 E., M.D.M., about 2 miles east of Pine Grove. Ownership: Benjamino Perini, Pine Grove, California.

During World War I, 30 tons of manganese ore was shipped from this property. In 1943 and 1944, Crabtree, Sullivan, and Stewart of Jackson shipped some ore to the Metals Reserve Company's stockpile in Auburn.

The manganese ore, chiefly psilomelane and pyrolusite, occurs as pods and lenses in massive metachert which is enclosed in a weathered quartz-sericite schist. The general structure strikes N. 23° and dips 65° northeast. In 1940, E. A. Kent was contracted to diamond drill this property. Although several stringers of rhodonite were encountered by seven 85-foot drill holes, no commercial ore bodies were discovered.

Mining workings consist of two small 15-foot, partially caved shafts and two large bench cuts as well as extensive surface bulldozing.

Stirnaman Mine. Location: SE ¼ of sec. 24, T. 7 N., R. 12 E., M.D.M., between Pioneer Station and Pine Grove. Ownership: J. L. and L. C. Wells, 716 Wagner Avenue, Stockton, California.

During World War I, 250 tons of manganese ore were shipped from this property. Dr. J. T. Stacy of Pine Grove leased and mined the property during 1942 and 1943 and shipped some ore to the Metals Reserve Company in Sacramento. This property has been idle since 1943.

The manganese ore, chiefly psilomelane, occurs as lense 1 to 8 feet wide in a belt 30 to 50 feet wide of massive metachert which is enclosed by quartz-sericite schist. The general strike is N. 30° E. and the dip is about 60° southeast. One shipment during the fall of 1942 ran 52 percent manganese, 3 percent iron, and 4.5 percent silica (Trask, 1950, p. 35).

Development work consists of some small and large open cuts, two main levels of stopes, and surface bulldozing.

Platinum Group Metals

From 1904 to 1923 the American Dredging Company, which operated three gold dredges on the Mokelumne River, produced from 20 to 25 ounces of platinum group metals per year (Logan, 1927, p. 201). From 1923 to 1942 minor amounts were produced intermittently as a by-product of gold-dredging operations. Of the metal produced, the platinum content was about 50 percent and the remainder was chiefly osmiridium (Logan, 1927, p. 201).

Platinum occurs alloyed with related precious metals in smooth thin flakes or as rounded nuggets in placer deposits in the Sierra foothills. The proportion of platinum metals to gold is nearly always small. Primary platinum is believed to be deposited by magmatic processes in basic or ultrabasic rocks.

NON-METALLIC MINERALS

Asbestos

There are two distinct varieties of asbestos: chrysotile, a variety of serpentine, and amphibole asbestos, which includes several fibrous minerals such as tremolite and actinolite. In Amador County both amphibole and chrysotile asbestos occur in or near massive serpentine bodies.

In Amador County 10 tons of asbestos was produced in 1908 and 2 tons was produced in 1909. Since that date there has been no recorded production, although some intermittent development work on several prospects has proceeded. At the present time, George Thomas of Jackson is developing an amphibole asbestos prospect in sec. 34, T. 6 N., R. 11 E., 1½ miles southeast of Jackson.

Amphibole asbestos occurs in irregular veins along a serpentine-amphibolite contact in the NE ¼ sec. 20, T. 6 N., R. 10 E., 2½ miles northeast of Ione (Logan, 1927, p. 134).

Coal

Amador County coal (lignite), although generally low grade in terms of fuel value as compared with bituminous and anthracite coal, commonly has properties which make it valuable for other purposes. In western Amador County, coal (lignite) is found solely in the Ione formation of middle Eocene age where it is associated with beds of clay and sand.

Lignite beds in the Ione region in Amador County were mined for local use early in the 1860's (Goodyear, 1877, p. 79). Intermittent production from this area continued until 1888, and from this time until 1902, substantial amounts of coal were produced. However, from 1902

FIGURE 15.　American Lignite Products Company montan wax extraction plant, Buena Vista. *Photo by Mort D. Turner.*

FIGURE 16.　American Lignite Products Company open-pit coal mine. Coal bed is about 12 feet thick. *Photo by Mort D. Turner.*

until 1947, there was only sporadic production of coal due to competition from petroleum.

It was not until 1947 that coal mining in this county began again, and then not for fuel purposes. In the late 1940's the lignite was found to contain a high percentage of wax suitable for industrial purposes providing an economic means of extraction could be developed. Profitable extraction methods have since been developed and montan wax and other industrial waxes have been produced. Now Amador County deposits make the United States self-sufficient in montan wax (Sawyer, 1949), which was formerly imported from Germany. Montan wax is similar to carnauba wax and is used in carbon paper, phonograph records, polishes, and rubber.

American Lignite Products Company (Western Division of De Angelis Coal Company of Carbondale, Pennsylvania). Location: The plant is located approximately 6 miles south of Ione, near Buena Vista, in NE ¼ of sec. 19, T. 5 N., R. 10 E., M.D.M. The company owns 1800 acres in Ione including the Buena Vista coal mine in E ½ sec. 19, T. 5 N., R. 10 E., M.D.M. J. N. De Angelis is plant superintendent of the company and F. J. De Angelis is the general manager.

Operation began in early 1947 with the erection of a pilot plant for the extraction of montan wax from lignite. The first commercial production occurred in April, 1948, after about a year's intensive experimental and research work. A serious fire in the fall of 1948 completely destroyed the pilot plant. A second pilot plant was completed in early 1949, only to be destroyed in 1949.

Construction of a larger extraction plant was completed during February of 1950. This plant averaged approximately 50,000 pounds each month until September of 1951 when a new process was installed. Work was completed during February of 1952. Different types of wax are produced, with a combined average of approximately 130,000 pounds per month, according to F. J. De Angelis.

Lignite occurs interbedded with the nearly flat-lying clay and sandstone beds of the Ione (Eocene) formation. According to the company's drill hole results, the lignite beds are lenticular-shaped bodies with a maximum diameter ranging from 500 to 3000 feet. The lignite seams range in thickness from about 1 foot at the outside edge of the lens to an approximate maximum thickness of 14 feet. The thickness of the overburden already encountered in stripping operations varies from 15 to 25 feet.

Strip-mining of the lignite beds is accomplished by an 85-foot boom dragline. The dragline bucket has a 2½-yard capacity and can handle about 800 tons of lignite per day. The conventional procedure used from strip mining is to recover the lignite along the edges of the deposit and work towards its center. End-dump trucks then transport the material to the extraction plant.

Lignite from the mine is conveyed to a hammer mill equipped with air separation equipment after having been crushed by the toothed crusher. The crushed material is fed into a hammer mill while the outgoing 200-mesh material is fed in weighed quantities to extractors where petroleum solvents remove waxes, organic salts (esters), and resinous and asphaltic substances.

FIGURE 17.   Strip mining at American Lignite Products Company pit.
Photo by Mort D. Turner.

The heavy material or "bottoms" from the centrifuge is directed into stills and finally to the dump. The light liquid or effluent passes through filters and into the distilling towers where montan wax is extracted by steam distillation. The wax is drained from the bottom of the still kettle into cooling pans, then to bags for shipment.

The plant is operated on a 24 hours per day, seven days per week schedule by 28 men. In addition, two chemists are employed in the company laboratory to insure quality control of the montan wax.

Humacid Company Coal Mine. Location: SW ¼ sec. 29, T. 5 N., R. 10 E., M.D.M., about 2½ miles south of Buena Vista in China Gulch. Ownership: The Humacid Company of Los Angeles has leased the property from B. F. Wellington, owner.

Although the lignite beds fluctuate locally somewhat in attitude, they occur essentially as flat-lying beds interbedded with Ione sands and clays. Small 1- to 2-inch clay beds occur in the lignite 2 to 3 feet apart. Samples cut by the company in the lignite beds revealed that the wax content is higher towards the bottom of the beds, the best material averaging 16 percent, according to W. C. Kreth, superintendent. Fossil plant remnants are a common occurrence at the tops of the beds. Preparatory to development and mining operations, four vertical test drill holes 200 feet apart on an east-west line were drilled to prove the dimensions of the lignite beds.

In 1951, the company began to sink a circular, vertical shaft 5 feet in diameter. The shaft, which is cased with a steel liner, is now 80 feet deep with one mining level entirely within the lignite at the 75-foot mark. On the 75-foot level, the operators have drifted 100 feet east, and then 100 feet north and 8 feet west, and then 100 feet north. Most of these underground workings are timbered. Continued drifting on the

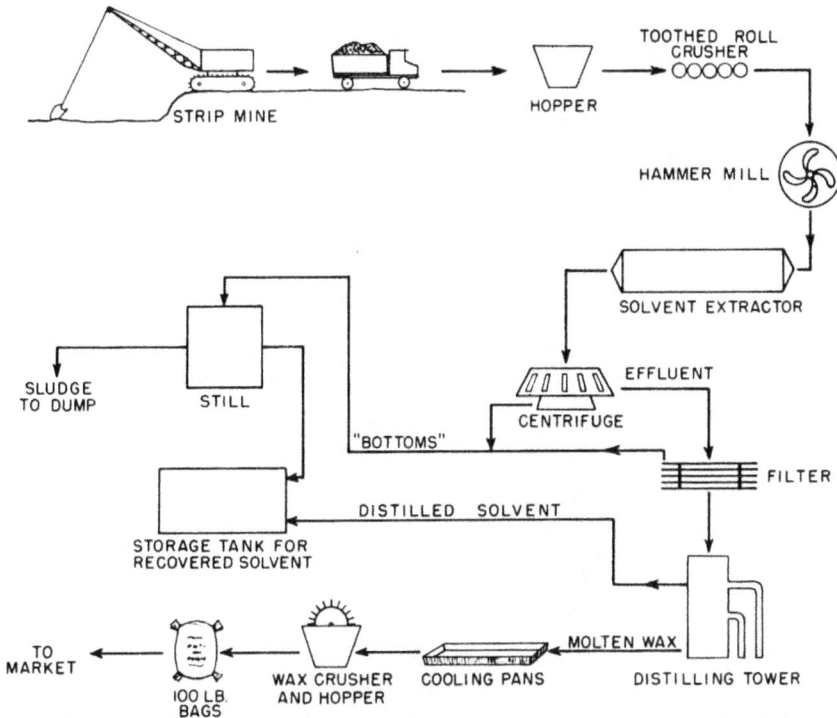

FIGURE 18.   Generalized flow sheet showing recovery of montan wax from lignite.
After American Lignite Products Company plant.

east drift and the north drift off the east drift is now underway. Future plans are to use the room-and-pillar method of mining.

The mine is lighted by incandescent lamps. A converted electric hoist powered by a Ford Model A gasoline engine lifts the circular cage in conjunction with a cement-mounted, 24-foot, welded steel headframe. Drilling is accomplished by four-foot, electric auger drills.

After blasting, the lignite is hand mucked into an ore car with an 1100-pound capacity, hand-trammed to the station, and loaded aboard the circular cage. The cage is then hoisted to the surface where the ore is again hand-trammed and the lignite deposited into a storage bin. The ore is then hauled about 3 miles to the company's plant by truck.

In late 1951, the mine produced 25 cars of ore per day for processing in the company's industrial wax plant. W. C. Kreth, superintendent, and two men operate the mine.

The Humacid Company is now in the process of erecting a plant to extract industrial waxes.

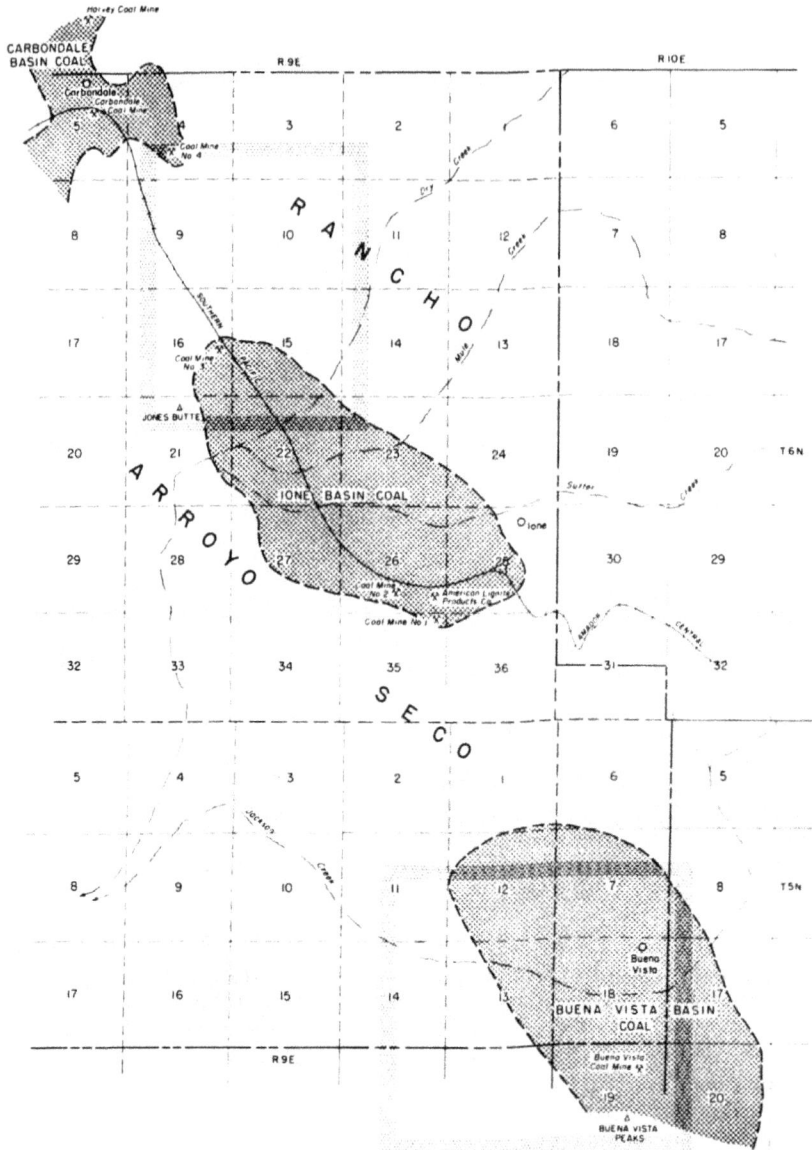

FIGURE 19.   The coal basins of Amador County. Boundaries are fairly definite for the east and west sides of the Buena Vista Basin, the south side of the Ione Basin and the portion shown of the Carbondale Basin. Other boundaries are approximately located. (Modified by Mort D. Turner from unpublished maps by Mocine, 1923, and McCready, 1927.) Areas within stippled borders are described in California Journal of Mines and Geology, July 1949, and in California Division of Mines Special Report 19.

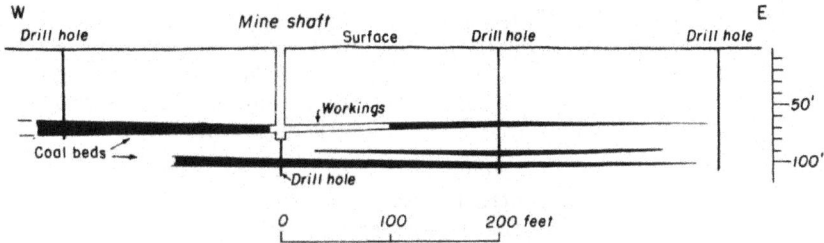

FIGURE 20.   Cross-section of Humacid Coal Company mine.

Crushed Rock, Sand and Gravel

Granodiorite, granite, and associated granitic rocks occur in the eastern part of Amador County. Broken and crushed granite is produced intermittently for use as road metal and fill. Coarse residual sand originating from the weathering of granite has been used in road repair and construction.

Riprap composed of broken granodiorite was used in the construction of the roads owned by the Pacific Gas and Electric Company east of the West Point Power House. In 1951 a portable crushing plant operated by A. R. Milton produced crushed granite for use as road metal by the Amador County Road Department. Coarse sand originating from weathered granite has been produced by the Amador County Road Department from a pit in sec. 1, T. 6 N., R. 11 E., M.D.M., on the New York Ranch.

Sand and gravel deposits are largely stream-laid accumulations of sand, gravel, cobbles, and boulders. The deposits are usually poorly-sorted, imperfectly stratified, and elongated in the direction of stream flow.

As the drainage pattern of Amador County is essentially westward and most of the rock formations strike northwest, its rivers, streams, and tributaries cut across many different rock types. The sand and gravel deposits are derived from rocks resistant to weathering. The resistant rock types consist chiefly of chert, metachert, greenstone, metavolcanic rocks, quartz, and granitic rocks. They are intermittently prepared and utilized as road metal and fill.

Exact production figures for sand and gravel in Amador County over the years are not available as this information is included in a miscellaneous category. However, during 1947, 34,490 tons of sand and gravel worth $65,476 was produced in the county (Jenkins, 1950, p. 42). Since that time, only small, intermittent production has been recorded.

Alpine Gravel Plant.   Location: NE ¼ of SW ¼ sec. 22, T. 6 N., R. 11 E., M.D.M., about 1 mile east of Jackson on Jackson Creek. Ownership: the plant is owned by Jack Bacigalupi, Pine Grove, California.

Jack Bacigalupi and Orville Youmans, both of Pine Grove, leased and worked the property from 1945 to June of 1951. During this time, the operators produced minus ¾-inch crushed rock for various contractors for use in concrete. The average production of the gravel plant was about 50 cubic yards per day, according to Mr. Bacigalupi.

Between 1946 and 1947, the operators recovered some gold by diverting the creek gravel through a dragline dredge before running it through the gravel plant. Jack Bacigalupi stated that the enterprise was not particularly successful as the gold content ran between 17 and 20 cents per cubic yard. The dredge is now idle.

Relvas Gravel Deposit. Floyd O. Bailey, general contractor of Madera, California, has a contract with Sacramento County to produce 1¼-inch road ballast from the old placer tailings on Dry Creek, about 2 miles northwest of Ione in secs. 14 and 15, T. 6 N., R. 9 E., M.D.M. (projected). The tailings piles are leased from E. J. Relvas, Ione, California. Three men operate a Twin-dual master 24 Universal portable crusher to produce about 1500 tons of crushed rock per day. The 1¼ inch material is stockpiled and used as needed by Sacramento County, Amador County, and the State of California.

Sacramento County Sand and Gravel Pit. Location: SW ¼ sec. 5, T. 7 N., R. 9 E., M.D.M., on Arkansas Creek just north of state highway 16, approximately 2 miles east of Michigan Bar. Ownership: Louis G., Cora J., and Gene L. Klotz, Freeport Boulevard, Route 8, Box 1380, Sacramento 17, California.

During 1951, this property was worked by the Sacramento County Highway Department. At that time, 7000 tons of crushed rock were produced from cobbles and coarse gravel and used as road metal. Sacramento County again commenced mining operations during the middle of June, 1952, producing sand. In the fall of 1952, 9000 tons of crushed rock were produced. This deposit is worked at irregular intervals to supply material for building and repairing roads in the eastern part of Sacramento County. The sand consists almost entirely of quartz.

Mining equipment includes a P & H gas-powered truck-mounted crane with a 30 foot boom and a ½-cubic yard clamshell bucket, a bulldozer, and several end-dump trucks.

The production of sand is intermittent depending upon the needs of the county. Two men operate the sand pit.

Dimension Stone

Rhyolite Tuff

Rhyolite, a fine-grained volcanic rock composed predominantly of light-colored feldspars and quartz, is occasionally utilized as dimensional stone. One such deposit in Amador County is the Evans deposit in the center of sec. 18, T. 7 N., R. 12 E., M.D.M., and extending to the northeast in sec. 17, T. 7 N., R. 12 E., M.D.M.

This material is a well-compacted rhyolite tuff varying in color from white to buff. During the 1880's and 1890's, rhyolite from this deposit was quarried as a building stone for various buildings in the county, according to Elmer Evans, Jr., the owner.

Sandstone

Appreciable quantities of sandstone were produced in Amador County prior to 1914. Since that date only minor amounts have been produced intermittently. This early production was largely dimension stone for use in building construction.

In the 1880's and 1890's, sandstone for dimension stone purposes was produced from a quarry in sec. 27, T. 5 N., R. 10 E., about 8 miles south of Ione. Even, fine-grained sandstone with a pleasing bright red color was produced from several quarry faces 15 to 20 feet high. Stone from this quarry was used in the construction of the Preston School of Industry in Ione, the California Bank Building in Sacramento, and the old Chronicle Building in San Francisco (Aubury, 1906, p. 117). Pure white sandstone having somewhat lower crushing strength than the red material may also be found in the quarries.

Serpentine

Small amounts of serpentine were produced in Amador County in the 1880's and 1890's for use as ornamental building stone. A yellowish-green to dark olive-green serpentine was quarried 2 miles west of Plymouth while a mottled variety of serpentine was quarried 1½ miles west of Sugar Loaf in the vicinity of Waits Station (Aubury, 1906, p. 147).

Slate

In 1941, G. J. Alexander of Amador City produced slate from a quarry near Martell. About 1900, some preliminary work was done on a slate quarry 1½ miles east of Ione. Except for these operations, there has been little production.

Slate is used primarily as roofing granules and slabs, flagstone, table tops, blackboards, insulating material, and as mineral filler (Turner, 1950, p. 259). Most of the slate in Amador County is found in the Mariposa formation, which occurs in broad belts in the western part of the county.

Gems

Diamonds have been found in gold-placers in the Sierra Nevada foothills in Amador County. As yet no primary deposits have been discovered but the placer diamonds are believed to have originated in the ultra-basic igneous rocks from which the serpentine masses were derived.

Between 60 and 70 small diamonds have been found in Jackass Gulch near Volcano. The larger stones ranged from 1 to 1.57 carats in weight (Murdock and Webb, 1948, p. 30). Four small octahedral diamond crystals found near Volcano were placed on display in the American Museum of Natural History, New York City (Kunz, 1905, p. 41). A few small diamonds have also been found in Indian Gulch near Fiddletown. In 1934, a 2.65 carat diamond was found near Plymouth (Sperisen, 1938, p. 39).

Several varieties of chalcedony and opal occur in Amador County, although there is no recorded production of these commodities. Chrysoprase (apple-green chalcedony), occurs as seams in altered serpentine 6 miles southeast of Ione in sec. 34, T. 6 N., R. 10 E., M.D.M. Chrysoprase also occurs in veins and as replacement pods in silicified serpentine 1 mile south of River Pines east of the Pigeon Creek School. Small amounts of blue chalcedony occur in the vicinity of Volcano (Pabst, 1938, p. 96).

Moss opal is found three-quarters of a mile south of River Pines associated with siliceous limonitic serpentine. Opalized weed has been found near Volcano (Pabst, 1938, p. 99).

Limestone-Marble

Lime (CaO) and limestone ($CaCO_3$ + impurities), two closely allied materials, are the basis of a multimillion dollar business in California

(Bowen, 1950, p. 171). Most lime is produced by calcining limestone. The largest consumer of lime is the portland cement industry which requires a high-calcium limestone. Much lime is also used in mortar, plaster, and stucco.

Other important uses of limestone are in agriculture as soil additives, in the steel industry as a basic flux in furnaces, in the beet sugar

FIGURE 21.   Old building in Volcano, constructed of limestone quarried locally. *Photo by Mary Rae Hill.*

industry to remove impurities from the raw beet juice, and in the manufacture of industrial chemicals.

Total production of lime and limestone from Amador County has not been recorded (Logan, 1947, p. 207). Between 1904 and 1915 production of lime averaged 1100 barrels per year. Since 1915 little or no limestone has been produced except in 1945 from the Allen property.

In 1894 almost 26,000 cubic feet of marble were produced. From 1895 to 1905 marble production averaged slightly more than 4000 cubic feet per year. Since 1905 only small amounts of marble have been produced intermittently from the Dondero quarry.

Limestone lenses in Amador County occur in the Calaveras formation in the western Sierran foothills and farther east near Volcano and Fiddletown. In the west the lenses are numerous but small in areal extent while to the east they are considerably larger, the largest being the Volcano deposit.

The limestone has a north to northwest trend and occurs in lensoid to tabular bodies which stand nearly vertical. Jointing is prominent in nearly all of the deposits. Those in the western foothills and at Volcano are bluish-gray in color and weather to a dull gray. Several of the deposits consist of marble that is white with irregular blue and gray streaks. West of Drytown is a deposit of red marble. Crinoids and corals of Paleozoic age have been found in two lenses on the Allen property south of Sutter Creek.

A series of spot surface samples taken by O. E. Bowen, Jr., and C. R. Nichols at 20-foot intervals on the slope immediately north of Volcano and parallel to the main street was analyzed by Abbot A. Hanks, Inc., to give the following results by percentages:

	Sample Number							
	20	40	60	80	100	120	140	160
Insoluble	1.06	1.03	0.38	0.65	1.71	0.68	0.78	0.66
R₄O₃*	0.25	0.16	0.55	0.46	0.20	0.27	0.69	0.14
CaCO₃	97.40	97.23	97.06	97.23	95.39	97.21	96.70	96.96
MgCO₂	0.20	0.21	0.20	0.92	2.38	1.42	1.13	0.20

* Aluminum and iron oxide.

The following are analyses made by Abbot A. Hanks, Inc. of samples taken from various properties in Amador County:

Allen Estate, 7 miles northeast of Ione;
method of sampling unknown.

	Percent
Insoluble	1.60
Ferric and aluminic oxides	0.42
CaCO₃	96.90
MgCO₃	1.06

Garibaldi Ranch, 2 miles south of Drytown; composite sample
across width of 120 feet by C. A. Logan.

	Percent
Insoluble	3.55
Ferric and aluminic oxides	0.30
CaCO₃	71.70
MgCO₃	24.31

Dal Porto Marble, 2 miles east of Plymouth; composite
sample across quarry face by C. A. Logan.

	Percent
Insoluble	5.59
Ferric and aluminic oxides	0.56
CaCO₃	90.69
MgCO₃	2.04

Fiddletown deposit; composite surface
sample by C. A. Logan.

	Percent
Insoluble	0.60
Ferric and aluminic oxides	0.22
CaCO₃	95.31
MgCO₃	2.98

Grelich Ranch, 8 miles southeast Latrobe;
method of sampling unknown.

	Percent
Insoluble	1.00
Fe₂O₃ and Al₂O₃	0.56
CaCO₃	97.85
MgCO₃	0.41

Allen Limestone Deposit, Teichert Lease. Location: SW¼ sec. 13, T. 6 N., R. 10 E., M.D.M., 2½ miles southwest of Sutter Creek. Ownership: George Allen, Sutter Creek.

A. Teichert and Son leased this property in 1945 and shipped 5000 tons of limestone for use as sugar rock. The fines were sold as road rock. This property has been idle since 1945, according to Lee Gardner.

The deposit consists of a limestone lens approximately 100 feet wide and 400 feet long. The limestone is bluish-gray, fossiliferous, prominently jointed, and strikes northwest.

The limestone was handled by power shovels, crushed, sized by a multi-deck trommel, and shipped by truck. There was no equipment on the property late in 1952.

Dondero (Carrara) Marble Quarry. Location: N½ sec. 29, T. 7 N., R. 12 E., M.D.M., 2¼ miles west of Volcano. Ownership: Aurelio G. Dondero, 852 59th Street, Oakland, California.

This property was worked on a modest scale for 30 years until 1933. A quarry floor 150 feet wide with a face nearly 150 feet high was worked (Logan, 1947, p. 209). Large blocks of marble in a rough state were shipped to San Francisco where they were cut and polished. Marble from this quarry was used in the rotunda of the San Francisco City Hall and in the rotunda of the Museum Building at Stanford University (Aubury, 1906, p. 97).

The stone is principally white marble with black streaks, although white and ash-colored marble are also present.

Mineral Paint

In 1918 and 1920 a small production of mineral paint was recorded in Amador County. At many places in the Ione sands and clays, red and yellow iron oxides occur in the form of mineral paint or ochre. Because of impurities and physical properties of most of the deposits, very little ochre has been produced in Amador County.

Roberts Lease. Location: NE¼ sec. 34, T. 6 N., R. 11 E., 1½ miles southeast of Jackson. Arthur Roberts of Jackson leased the property during World War I for a short period on land that is now part of the Ginocchio estate, according to J. Schweitzer.

The raw material is highly weathered schist. This rock is very friable, fine grained, and some is streaked with white and yellow material. The deposit measures about a half a mile along the strike by about 150 feet wide.

Two carloads of ochre were shipped to the market during World War I, according to Mr. Schweitzer.

Pumice

Fragmental pumice, a frothy volcanic glass, is mined for aggregate at several points in Amador County.

Pumice and a fine-grained pumicite crushed from it are utilized principally as a lightweight aggregate in the construction industry. Other uses are as abrasives, as fillers in paints, paper, and rubber, insecticide carriers, and as filtering aid.

A small production of crushed pumice from Amador County has been recorded each year since 1938, with the exception of the war years, 1942 to 1946.

In 1947 Charles Bacon of Ione began sporadic production of pumice from a deposit in SE¼ sec. 19, T. 6 N., R. 9 E., M.D.M. (projected). This material is marketed principally in Berkeley, California.

This deposit, which occurs in the Valley Springs formation, consists of a bed of white pumicite 10 to 15 feet thick. The bed has a slight westerly dip and is overlain by andesitic detritus of the Mehrten formation. Overburden is absent in places, but may attain 90 feet in thickness in others. Development work consists of an open pit and a short adit.

Quartz Crystals

During World War II the U. S. Geological Survey conducted a quartz-crystal investigation in California. Although there has been no recorded production of quartz crystals thus far from Amador County, the investigation revealed that one of the most promising deposits in California [13] is located near Fiddletown. The crystals are found in

[13] Durell, Cordell, personal communication, 1952.

placer deposits between Dry Creek and Big Indian Creek, northeast of Fiddletown. Sustained economic exploitation of this deposit probably would be attempted only in times of national emergency. During World War II quartz crystals for use in vacuum tube oscillators were in short supply and all known resources were investigated.

The quartz crystal deposits near Fiddletown occur in ancient stream gravels, possibly of upper Eocene age. Most of the crystals are located in close proximity to the bedrock and show very little if any abrasion, indicating that they were transported only a short distance from their source.

Refractory Materials

Clay *

The clay minerals are comprised of three major mineralogic groups, the kaolins, the montmorillonites, and the illites. Kaolin-group minerals are kaolinite, dickite, halloysite, and nacrite, all of which have the ideal formula $Al_2Si_2O_5(OH)_4$. They are distinguished by their internal crystalline structure. Kaolinite is almost always the dominant mineral in commercial deposits of high-grade ceramic clays. Clays of the kaolinite group make up all of the commercial clay deposits of Amador County.

Clays are divisible into two main geological groups; residual (formed in their present position) and sedimentary (transported to their present position from a point of origin elsewhere). Most residual clays are formed by the weathering of materials at or near the surface or, less commonly, by alteration of rock by hot rising mineralized water. The sedimentary clays are derived from residual clays by erosion, usually by running water.

Most of the high-grade fire clays of California are in sedimentary deposits formed in the Paleocene and Eocene epochs. At this time topography was subdued, a large part of the state was covered by shallow seas, and a warm, moist climate prevailed. These conditions caused deep weathering of the land surface and led to the formation of residual clay over large areas. The clays were gradually eroded and deposited in lagoons and deltas. Beds that contain clay deposits of commercial interest are exposed in (1) a narrow, discontinuous belt along the western Sierra Nevada foothills; (2) a horseshoe-shaped belt on the west, north, and east sides of the northern portion of the Santa Ana Mountains, Riverside and Orange Counties; and (3) an area in the northern portion of the Diablo Range, Alameda and Contra Costa Counties. The important production centers are at Alberhill in Riverside County, Ione in Amador County, and Lincoln in Placer County. Amador County ranks third among the counties of California in total fire-clay production and in annual fire-clay production.

The fire-clay deposits of Amador County are contained in the Eocene Ione formation which formed in the shallow sea that bordered the western flank of the ancestral Sierra Nevada. The clay bodies are lenticular and nearly horizontal. Some of the clay-bearing beds have been capped and preserved from erosion by volcanic rock, principally rhyolite and andesite. The Ione formation in Amador County is composed of two members separated by an unconformity (Pask and Turner, 1952).

---

* This section has been prepared by Mort D. Turner, Assistant Mining Geologist, California Division of Mines.

The lower member of the Ione formation contains most of the commercial clay of the area and is characterized by the rarity of chlorite, biotite, and certain types of clays. In southwestern Amador County, the lower member is divisible into three lentils. The lower lentil contains the Edwin clay—which is mined near the town of Ione—and reworked laterite. The middle lentil contains the lignitic coal beds of the area. The upper lentil contains the Cheney Hill clays and the white Ione sand.

The upper member of the Ione formation is predominantly sandy and contains two mappable units: a hard white sandstone at the top of the member, and the Chitwood clay in the upper part of the member.

The most significant fact so far derived from the geological studies of the Division of Mines in the Ione area is that with only one exception (the Chitwood clay) all of the commercially valuable clays are restricted to the lower member of the Ione formation and to the weathered rock at the base of the Ione. This is of primary importance to any future prospecting for clay in the area. Many useful articles have been published on the geology of the clay-bearing beds of the Ione area and should be consulted by anyone working with clay in Amador County (see Pask and Turner, 1952).

Clay mining in Amador County began at least as early as 1864 with the opening of the Dosch Pit between Ione and Carbondale. However, other pits probably were already active in the Carbondale region by that time. As in other parts of the state, clay was discovered through mining for coal. With coal as a cheap plentiful fuel and good clay readily available it is not surprising that several prosperous potteries were built. At least one was active by 1862 and some were still producing as late as 1906.

As the use of refractory clay gradually overshadowed pottery clay the center of production moved toward Ione until at the present time there is only one active clay pit near Carbondale; all the rest are in the Ione Valley. While a great deal of the Ione clay still goes into heavy clay products, a large part, probably well over half, is used as refractory material. Especially valuable to the economy of the state is the presence of large quantities of Edwin clay in the county. This is a very refractory clay with a P.C.E. * of cone 34 + and therefore suitable for the production of the highest grade of fire-clay bricks.

The following descriptive listing of clay pits only covers those pits which are active now or have been active recently. A much more complete listing is contained in the tabulated list of mineral deposits although this list does not contain some of the old and small pits of which no record has survived.

Airplane Pit. Location: 3 airline miles N. 52° W. of Ione on the north side of the Ione-Sacramento road. The Airplane pit is in lots 222 and 236 of Rancho Arroyo Seco; NW¼ sec. 15, T. 6 N., R. 9 E., M.D.M. (projected). Ownership: Charles S. Howard Estate, c/o Crocker First National Bank of San Francisco, Post and Montgomery Streets, San Francisco, leased since 1948 to Gladding, McBean and Company, 2901 Los Feliz Boulevard, Los Angeles 26.

---

* P.C.E. (pyrometric cone equivalent) is a means of designating the refractoriness of a clay by comparison, during firing, to the behavior of a standard set of prepared pyrometric cones. The number is that of the standard pyrometric cone whose tip touches the support simultaneously with the tip of a cone of the refractory material being tested. The testing is done under standardized conditions.

Refractory clay is mined from the upper part of the lower member of the Ione formation. Seven beds have been recognized and called Bacon No. 1 to No. 7. Bacon No. 1 is the thin overburden of soil and contaminated clay. Bacon No. 2, which was formerly produced, has now practically disappeared. Bacon No. 3, a white clay, and Bacon No. 4, a chocolate-colored clay, are mined together and have a combined thickness of 8 to 10 feet. Bacon No. 5 and No. 6 are a light buff color with slightly different appearance but the same physical properties and are mined together. They have a combined thickness of 12 to 15 feet. Bacon No. 7 contains a much higher proportion of iron than the beds above and very little has been mined. Below Bacon No. 7 is a gray-blue sandy clay. Bates (1945, p. 24) classifies the Bacon clays as Eastern Ione type. The general distribution of these clays to the northwest is indicated by Johnson (1948 and 1949).

Clay has been produced from the Airplane pit since about 1930, and during this period the pit has been one of the major sources of refractory clay in the Ione area. Most of the clay produced is shipped by rail to Pittsburg for use in refractories and a small amount is used at Lincoln in heavy clay products. The deposit has been mined entirely by open pit. Stripping and mining are carried out under contract as described under Gladding, McBean and Company.

Cheney Hill Pit. Location: 2 airline miles S. 45° W. of Ione on the north side of Cheney Hill. The Cheney Hill pit is in lots 257 and 258 of Rancho Arroyo Seco; SW¼ sec. 35, T. 6 N., R. 9 E., M.D.M. (projected). Ownership: Charles S. Howard Estate, c/o Crocker First National Bank of San Francisco, Post and Montgomery Streets, San Francisco, leased since 1948 to Gladding, McBean and Company, 2901 Los Feliz Boulevard, Los Angeles 26.

Cheney Hill clay, one of the several type-clays that crop out over wide areas of the Ione Valley, has been mined at a number of places. The bluish-gray Cheney Hill clay occurs in the upper part of the lower member of the Ione formation. In this pit it has a thickness of about 17 feet, is overlain by as much as 24 feet of brown clayey sand and sandy clay, and is underlain by more clayey sand. The Cheney Hill clay has a P.C.E. of cone 33-34. (Bates, 1945, p. 24.)

The Cheney Hill pit has been worked since the late 1920's and the clay produced used largely in the manufacture of refractories. The deposit has been mined entirely by open pit. Stripping and mining are carried out under contract as described under Gladding, McBean and Company.

Clark Sand Pit. Location: 1.7 airline miles N. 16° E. of Carbondale; SW¼ sec. 28, T. 7 N., R. 9.E., M.D.M. Ownership: Pacific Clay Products Company, Box 145, Station A, Los Angeles 31.

Clayey sand is mined from the Ione Sand bed near the top of the lower member of the Ione formation. The usable material is from 25 to 40 feet thick and is overlain by soil and iron-stained sand up to 4 feet thick. The Ione Sand in this pit contains a higher proportion of sand to clay than is usual in Amador County.

The Clark sand pit was active at least as early as 1906 and there has been intermittent production to the present (Aubury, 1906, p. 208; Tucker, 1914, p. 6; Logan, 1927, p. 136; Dietrich, 1928, p. 261). In the

FIGURE 22.   Custer clay pit. West face in bank of siliceous Ione sand and gravel.
Photo by Mort D. Turner.

FIGURE 23.   Deutschke pit showing face of Edwin clay. Camera bearing northwest.
Photo by Mort D. Turner.

early 1920's a pilot plant was constructed for the separation of china clay and silica sand. The Ione sand from the Clark pit is used largely for setting sand with small amounts going into terra cotta and white bricks.

The earliest production was from room and pillar stopes but in recent years all operations have been by stripping and open pit operations.

Deutschke Pit. Location: 3.3 airline miles N. 75° W. of Ione; N½ sec. 21, T. 6 N., R. 9 E., M.D.M. (projected). Ownership: Charles J. Deutschke, Ione, under lease to Western Refractories Company, Russ Building, San Francisco.

Edwin clay is mined from the Edwin clay bed in the lower part of the lower member of the Ione formation. The bed at this point is about 20 feet thick with very little overburden. The clay is pink to white with a P.C.E. of cone 34 to 34 +. Some red spots, due to presence of iron, have a slightly lower P.C.E.

The Deutschke pit is stripped and then mined by open pit methods. The clay is trucked to the Western Refractories Plant at Ione where it is used in the production of super-duty fire bricks.

Dosch Pit. Location: 2.8 miles N. 52° W. of Ione at the intersection of the Sacramento and Irish Hill roads; N½ sec. 15, T. 6 N., R. 9 E., M.D.M. (projected). Ownership: Pacific Clay Products Company, Box 145, Station A, Los Angeles 31.

Flat-lying beds of white to buff refractory clays are mined from the upper part of the lower member of the Ione formation. These clays are similar to the Bacon clays of the adjacent Airplane pit but exact correlation is not readily possible. Two to 4 feet of soil and gravel overlie 10 to 12 feet of Dosch stripping, a clay similar to Bacon No. 2. The Dosch clay below the Dosch stripping is 50 to 60 feet thick, of which the top 20 feet is currently being mined.

The Dosch pit has been worked since 1864 and is the oldest continuously active clay mine in the state. It was operated by N. Clark and Sons of Sacramento, later of Alameda, until 1946. The clay is mined from an open pit by power shovel and trucked to sheds at Clarksona. From Clarksona it is shipped by rail to Stockton and elsewhere for use in a variety of ceramic products.

For references concerning the Dosch pit, consult the tabulated list at the end of this report.

Edwin Deposits. Location: 3.5 miles N. 72° W. of Ione around the base of Jones (Edwin) Butte. The deposits are in lot 240 of the Rancho Arroyo Seco; S½ sec. 16, N½ sec. 21, T. 6 N., R. 9 E., M.D.M. (projected). Ownership: Charles S. Howard Estate, c/o Crocker First National Bank of San Francisco, Post and Montgomery Streets, San Francisco, leased since 1944 to Gladding, McBean and Company, 2901 Los Feliz Boulevard, Los Angeles 26.

Highly refractory Edwin clay has been mined in four underground mines and four open pits from the Edwin clay bed. The Edwin clay is near the base of the lower member of the Ione formation and ranges up to 12 feet thick. Overlying the Edwin clay are two beds of lignite and a thick bed of Ione Sand. Below the clay is deep red clay derived from the erosion of older laterite. The Edwin clay is the most refractory of the Ione clays with a P.C.E. of cone 34 +.

FIGURE 24.   Dosch pit. The Dosch pit was opened in 1864 and is the oldest continuously active clay pit in California. Camera bearing north. *Photo by Mort D. Turner.*

The mining of Edwin clay at Jones Butte started at least as early as 1925 and has been continuous since. Only Edwin Pit No. 4 is now being worked. The clay is shipped by rail to Pittsburg where it is used for the manufacture of fire brick. Stripping and mining are carried out under contract as described under Gladding, McBean and Company.

For bibliographic references concerning the Edwin pit, see the tabulated list at the end of this report.

Farci Pit.   Location: 3.1 miles S. 27° E. of Ione; SW¼ sec. 5, T. 5 N., R. 10 E., M.D.M. Ownership: Antoni Farci, Ione, under lease since 1950 to Western Refractories Company, Russ Building, San Francisco.

White clayey sand is mined from the Ione sand bed in the upper part of the lower member of the Ione formation. Material from the Farci pit was used until 1949 in the production of white portland cement. Since then it has gone into the production of fire brick. The deposit was originally worked by room and pillar stoping but all of the later mining has been by open pit. In mining the overburden is stripped off and the Ione Sand loaded with a power shovel into trucks and hauled to the company plant at Ione.

Gladding, McBean and Company.   Gladding, McBean and Company, 2901 Los Feliz Boulevard, Los Angeles 26, Fred B. Ortman, President, operate several clay pits in the Ione area of Amador County and control a large number of idle or unexplored clay deposits. In 1948 they leased the clay-mining rights to the 33,276 acre Rancho Arroyo Seco

FIGURE 25.   Ione sand pit. Upper bench is cut in overburden, lower benches in white Ione sand. Camera bearing northwest. *Photo by Mort D. Turner.*

which covers most of the Ione-Carbondale clay area. In addition the Ione Red deposit is leased outside of the Rancho.

The various clay pits are stripped and mined under contract. The work usually begins in the spring, after the last heavy rains, with stripping of overburden from the clay bank to be mined that season. The carryall used for stripping is moved from one pit to the next until by the end of the stripping season all of the areas to be worked that year are uncovered. A crew with a bulldozer and a Hough mechanical loader progresses from pit to pit mining that year's supply of each particular clay in a few weeks. A caterpillar tractor with a rooter is used for either stripping or mining as the need arises. In this manner a large number of pits are worked each dry season with a minimum of men and equipment.

The clay is trucked to the nearest railroad siding and either loaded directly onto cars or stockpiled for future shipment. Gladding, McBean and Company do not use any of the Ione clays within the county, but ship them by rail to their plants in Lincoln and Pittsburg. In addition, large quantities are sold.

Ione Red Pit (Bacon Red) (Lane Mottled).   Location: 1.5 miles S. 43° W. of Ione on the south side of the Ione-Jackson road; NW¼ sec. 32, T. 6 N., R. 10 E., M.D.M. Ownership: Mrs M. J. Bacon, Ione. Leased in 1948 to Gladding, McBean and Company, 2901 Los Feliz Boulevard, Los Angeles 26.

A fine grained, plastic, red and buff mottled clay called the Ione Red or Bacon Red clay is mined from the upper part of the lower member of the Ione formation. The pit is on the extreme eastern edge of the Ione basin and the Ione Red clay rests directly on highly weathered Mariposa slate. The clay bed ranges from a few feet to 20 feet in thickness and the overburden in places is as thick as the clay bed. The beds have a shallow dip to the west.

The deposit has been worked as an open pit since about 1925. Stripping and mining are carried out under contract as described under Gladding, McBean and Company. The clay is hauled by truck to a nearby siding on the Amador Central railroad where a major proportion is shipped by rail to the Lincoln plant of Gladding, McBean and Company for use in heavy clay products.

Ione Sand Pit. Location: 1.3 miles S. 12° W. of Ione on the west side of the Ione-Stockton road. The Ione Sand pit is in lot 260 of Rancho Arroyo Seco; N½ sec. 36, T. 6 N., R. 9 E., M.D.M. (projected). Ownership: Charles S. Howard Estate, c/o Crocker First National Bank of San Francisco, Post and Montgomery Streets, San Francisco, leased since 1948 to Gladding, McBean and Company, 2901 Los Feliz Boulevard, Los Angeles 26.

At the Ione Sand pit Gladding, McBean and Company mine white clayey sand of the Ione Sand bed·in the upper part of the lower member of the Ione formation. The overburden is soil and iron-stained Ione sand up to 10 feet thick.

The Ione Sand pit was first operated about 1935 and has been a major source of fire sand since then. The fire sand is mined by stripping and open pit mining as described under Gladding, McBean and Company. It is then trucked to the railroad and hauled to Lincoln and elsewhere for use as setting sand and to a lesser extent in fire brick.

Kaolin-Frye Pit. Location: 0.5 of a mile N. 41° E. of Buena Vista on the east side of the Ione-Buena Vista road; SW¼ sec. 8, T. 5 N., R. 10 E., M.D.M. Ownership: The north end of the property is owned by the Calaveras Cement Company, 315 Montgomery Street, San Francisco 6, and the south end is leased by the Calaveras Cement Company from Pearl T. Fye.

A very low-iron phase of the Ione sand is mined from the Ione sand bed in the upper part of the lower member of the Ione formation. Reddish brown terrace gravel and alluvium up to 18 feet thick overlie white and yellow Ione sand up to 25 feet thick. Below the Ione sand is lignite and black clay. The pit is on the east side of the Buena Vista basin where the beds have a very low dip to the west.

The pit was started in January, 1950 and has been the source of a low-iron, high-alumina, high-silica material for the manufacture of white portland cement at the Calaveras Cement Company plant at San Andreas, Calaveras County. Only the white Ione sand is usable for this purpose, and therefore the yellow Ione sand is stripped off and rejected along with the reddish brown conglomerate above. The white clayey sand is loaded into trucks with a pay-loader and hauled to the plant at San Andreas. (See Pask and Turner, 1952.)

Laterite Pit. Location: 4.0 miles N. 67° W. of Ione in lot 238 of Rancho Arroyo Seco. The Laterite pit is in the W½ sec. 16, T. 6 N., R. 9 E., M.D.M. (projected) just north of Jones Butte. Ownership: Charles S. Howard Estate, c/o Crocker First National Bank of San

Francisco, Post and Montgomery Streets, San Francisco, leased since 1948 to Gladding, McBean and Company, 2901 Los Feliz Boulevard, Los Angeles 26.

Deep red laterite and lateritic clay are mined where they crop out a short distance north of the Edwin clay area around Jones Butte. The laterite was formed by intense tropical weathering during the Eocene epoch from Jurassic greenstone of the Logtown Ridge formation. The industrial value of the laterite is in its relatively low silica content (25 to 30 percent SiO_2) and mining is therefore restricted to the upper, more weathered part of the laterite profile where silica has been leached out. Downward the weathering becomes progressively less pronounced until fresh rock is reached. The entire weathered section is in many places over 100 feet thick but only the upper 10 to 15 feet have the desired chemical composition. Much of the laterite mentioned in the literature on the Ione area is not residual laterite but sedimentary laterite that was reworked into the base of the Ione formation. Such material immediately underlies the Edwin clay in most localities where the base of the Edwin clay has been exposed.

The presence of laterite in the Jones Butte area has been noted by most of the geologists working in the area (Whitney, 1865, p. 289; Turner, 1894; Dietrich, 1928, p. 54; Allen, 1929; Bates, 1945, pp. 13-15). It was not until 1949, however, that a commercial use was found for this laterite. In that year the Permanente Cement Company at Permanente, Santa Clara County, started using it in portland cement. The laterite is blasted and loaded into trucks with a power shovel, and then hauled to the railroad about half a mile away where it is shipped to Permanente.

Newman Red Pit. Location: 1.1 miles S. 18° E. of Ione in part of the old Newman pit. The Newman Red pit is on the east side of the Amador Central railroad and just north of the Western Refractories Company plant; NW¼ sec. 31, T. 6 N., R. 10 E., M.D.M. Ownership: Nettie V. Gradwohl (1945). The pit is leased by Western Refractories Company, Russ Building, San Francisco.

Red and buff mottled, plastic clay called Newman Red clay is mined from the upper part of the lower member of the Ione formation. It is very similar in its appearance and physical properties to the Ione Red clay except that it contains less sand and has a P.C.E. 2 to 3 cones higher. The Newman Red clay is 18 to 20 feet thick in this pit, with a bed 1 foot to 3 feet thick of white clay at the top and 10 to 15 feet of stripping consisting of clayey sand above that.

The pit was started in 1927 and has been active continuously since then. Mining is from an open pit with a power shovel after the sandy overburden has been removed. The thin bed of white clay above the red mottled clay is mined along with the Newman Red. None of the Newman Red is used in Western Refractories Company's Ione plant but is shipped to ceramic plants elsewhere.

For references dealing with the Newman Red pit, see the tabulated list at the end of this report.

Tiger Creek Deposits. Location: SW¼ SW¼ sec. 23, T. 8 N., R. 14 E., M.D.M., 1.6 airline miles due east of Cooks Station. Ownership: Winton Lumber Company, Martell, California.

FIGURE 26. Newman Red pit showing thin bed of white clay between overburden and Newman Red clay. Camera bearing southeast. *Photo by Mort D. Turner.*

FIGURE 27. Solari pit. Upper half of face is overburden and lower half is red and buff Ione sand. Camera bearing west. *Photo by Mort D. Turner.*

This deposit of halloysitic clay, which is exposed by Tiger Creek and a tributary creek, is approximately 500 feet in diameter. The clay is brittle with low plasticity, and the color is white, pink, and brown to rust-red. Overburden consists of alluvium and soil as much as 10 feet in thickness. Iron oxide is concentrated irregularly through the clay body, while quartz and other mineral grains are present as impurities. This deposit is the result of hydrothermal alteration of andesitic sediments and agglomerates of the Miocene Mehrten formation.

There is no recorded production of clay from this property although there is a 17-foot vertical shaft.

Western Refractories Company. Western Refractories Company, Russ Building, San Francisco, operates a number of clay pits and a refractory brick plant in Amador County. The various pits are separately described.

The Western Refractories plant (Tucker, 1914, p. 11; Boalich, 1920, p. 38; Logan, 1923, p. 96; 1927, p. 141; Dietrich, 1928, pp. 60-61) is 1.4 miles S. 18° E. of Ione on the Amador Central railroad; NW¼ sec. 31, T. 5 N., R. 10 E., M.D.M. The plant was built in 1906 by the Ione Fire Brick Company and purchased by the Western Refractories Company in 1944.

Raw materials passing through the plant follow three different flow patterns. One for stiff-mud brick, another for dry-press brick, and a third for ground fire clay. For stiff-mud bricks the various clays and grog are fed into a dry pan and ground together. After screening, the batch is mixed with water in a pug mill, passed through a de-airing machine, and extruded as a bar. The bar is wire cut and the bricks repressed. These green bricks are dried in the drying yard or pass through a dryer. After drying they are fired. Clays for dry-press bricks are separately ground in a dry pan and then mixed in a wet pan where a small amount of water is added. The mix is sent to a dry press and from there directly to the kilns for firing. Clay to be sold as ground fire clay is ground, screened, and bagged.

The plant contains six bee-hive, down-draft kilns fired by fuel oil. The kiln cycle is as follows:

setting ----------------------------------	4 days
burning ----------------------------------	11 days
cooling ----------------------------------	8 days
drawing ----------------------------------	4 days

It therefore takes 27 days from the time the loading of a kiln is started until it is ready to load for the next cycle.

Seven grades of fire bricks, with various shapes of each, are produced by the Western Refractories Company at its Ione plant.

Yosemite Pits (Harvey Pits). Location: 1.2 miles N. 16° W. of Carbondale; W½ sec. 29, T. 7 N., R. 9 E., M.D.M. Ownership: Western Refractories Company, Russ Building, San Francisco.

Bluish-gray fire clay is mined from the lower member of the Ione formation. There are two types, Yosemite No. 1 clay, which is in beds 8 to 9 feet thick and has a P.C.E. of cone 32 to 33, and Yosemite No. 2 clay, in beds below the No. 1, which is 6 to 10 feet thick, more sandy, and has a P.C.E. of cone 30 to 31.

Above the Yosemite clay is 5 to 5½ feet of reddish brown gravelly soil and below is carbonaceous clay and lignite and then more clay. The clay beds gradually thin to the north.

The Yosemite clay is used primarily in refractories. It is mined in large shallow pits by power shovel and hauled by truck to the Western Refractories plant at Ione. There are three pits on the property, two of which are active.

FIGURE 28.   East face of Yosemite pit showing bank of Yosemite No. 1 clay.
Photo by Mort D. Turner.

Glass Sand

From 1903 to the middle 1920's there was intermittent production of glass sand in Amador County. All of the glass sand now produced in California is used in the manufacture of bottle ware, although some early production from Amador County went into window glass (Turner, 1950, p. 257). The sands occur interbedded with clay in the Ione formation.

Prior to and during World War I and in the early 1920's, glass sand was produced from a pit half a mile east of Ione on the Southern Pacific Railroad. It was operated by the Philadelphia Quartz Company in the early 1920's. Sand and clay were mined, broken up in heaters, and sent to classifiers, which separated the sand from the clay. The sand was then run over concentrators, which removed the heavy black sand, and the glass sand was recovered as middling. Production was 25 tons of sand per day (Hamilton, 1920, p. 39).

Custer (Grog) Pit. Location: 1.3 airline miles S. 52° E. of Ione: SE¼ sec. 30, T. 6 N., R. 10 E., M.D.M. Ownership: E. W. Custer and leased to Western Refractories Company, Russ Building, San Francisco.

Sand and gravel, with a silica (SiO_2) content of over 90 percent, is mined from a gravel bed in the lower(?) member of the Ione formation. The portion of the bed that is mined is over 20 feet thick and lies under 10 to 12 feet of overburden.

The pit that is now being worked was begun about 1944 but adjacent pits in the same bed are much older. Mining is in an open pit by means of a power shovel. The sand and gravel is trucked to the Western Refractories Company plant where it is crushed and used as a grog with refractory clay. The pit is active at the present time.

Soapstone and Talc

Small amounts of soapstone have been produced intermittently from Amador County since 1911. Although soapstone commonly contains a higher proportion of the mineral talc (basic magnesium silicate) than do some commercial "talcs," the term soapstone, as ordinarily used, implies the presence of impurities that prevent the use of the material as a high-grade commercial talc (Wright, 1950, p. 277). Most of the soapstone produced in Amador County has been utilized as a roofing filler, soapstone pencils, and for other minor industrial purposes. Soapstone deposits are most commonly associated with serpentine and talc-actinolite schist.

Light green talc crops out 6 miles northeast of Ione in sec. 4, T. 6 N., R. 10 E., M.D.M. In 1939, the Walter S. McLean Company of San Francisco did some work on a soapstone deposit east of Ione. D. E. Jones of Pine Grove produced soapstone during 1944.

MINERAL PROCESSING

Calaveras Ironstone Company. Location: NW¼ sec. 11, T. 7 N., R. 10 E., M.D.M., at the site of the old Plymouth Consolidated mine yard, and subsequently, the site of the Plymouth Rock Wool Manufacturing Company's plant in Plymouth.

In August, 1950, John Johnson, self-employed owner of the company, began to screen, clean, and bag tiny glass beads, formerly a waste product from the manufacture of rock wool, for use in sandblasting. Three sizes of 7, 20, and 30 mesh beads are bagged and shipped to the market. Production is intermittent, depending upon the market conditions, according to Mr. Johnson.

Dunlavy Concrete Block Plant. Location: ¼ mile south of Jackson on old state highway 49.

In 1946 Leo Dunlavy erected a plant for the manufacture of concrete blocks and has operated almost continuously since that date. Pumice, shipped by railroad from Bend, Oregon, sand from Lancha Plana, and Portland cement supplied by the Pacific Portland Cement Company are the principal constituents used.

By volume, two-thirds pumice and one-third sand are mixed to form an aggregate. One part cement is then added to 8¾ parts aggregate, 6½ gallons of water are added per cubic foot of cement used. Three ounces of D-4 granular soap are added to increase the plasticity of the mixture and a small amount of coloring matter is also added. This material is mixed in an electrically operated pugmixer having a capacity of 9 cubic feet. Mr. Dunlavy reports that the raw material is mixed for 15

FIGURE 29.   Dunlavy concrete block plant, Jackson. Concrete blocks are
drying in right background.

minutes in the pubmixer which rotates at a rate of 26 revolutions per minute.

The mixture is then delivered by a 21-foot rubber conveyor belt into a hopper. From the hopper it is introduced by gravity into two block molds mounted on an electrically operated vibrator. Each mold shapes four concrete blocks at one time. The vibrator causes the plastic mixture to fill the mold completely. The wet blocks are then pressed out of the molds by a press, operated by hand lever, onto wooden pallets.

The pallets, each with four blocks, are wheeled into a yard where the blocks are stacked for drying. Complete drying of the blocks requires at least 28 days. The blocks range from light gray to light pink in color. The most popular size is 8-by-4-by-16 inches.

Concrete blocks are used chiefly in the construction of private homes although they find some use in the construction of small public and farm buildings.

Concrete blocks are sold on a contract basis to the consumer so that enough blocks are produced at one time to build a complete structure. An average five-room house requires between 1400 and 1800 8-by-4-by-16 inch blocks. Two or three men including Mr. Dunlavy operate the plant.

Orr Custom Mill.   Location: NW¼ sec. 10, T. 7 N., R. 10 E., M.D.M., 1¼ miles west of Plymouth just south of the Plymouth-Latrobe road. Ownership: Orr, Ellwood, Plymouth, California.

Early in 1936 Ellwood Orr of Plymouth erected a mill which catered to small gold operators. The mill was in operation until the middle of 1937, according to Mr. Orr. Since that date it has been idle.

Ore was delivered by truck and dumped into a bin. From there it went through a 10-stamp mill, amalgamation tables, a Wilfley table, and Frue vanners. A small ball mill was used for regrinding. Capacity was 50 tons per day. When in operation six men were employed.

Plymouth Rock Wool Company. Location: Near the center of sec. 11, T. 7 N., R. 10 E., M.D.M., on Plymouth Consolidated mine property. W. H. Danzer of North Sacramento leased the property from the Argonaut Mining Company, owners of the property at that time. The manufacture of rock wool for insulating purposes began in 1944 and continued almost steadily until the fall of 1950, according to Lawrence Burke. The plant has since burned down.

Copper slag from the Copper Hill mine was mixed with coke and melted in a brick-lined furnace. Pressurized steam was blown through the molten mass to form a fluffy rock wool. The finished product was bagged and shipped to market. On construction jobs requiring insulation, a portable blower was mounted on a truck and rock wool was delivered into the area to be insulated.

Plant capacity ranged from 4000 to 8000 pounds of rock wool per day. During the years of operation the plant utilized 8000 to 10,000 tons of slag per year, according to Mr. Burke.

BIBLIOGRAPHY

Allen, V. T., 1928, Anauxite from the Ione formation of California: Am. Mineral, vol. 13, pp. 145-152.

Allen, V. T., 1929, The Ione formation of California: Univ. California Dept. Geol. Sci. Bull., vol. 18, pp. 347-448.

Aubury, Lewis E., 1905, The copper resources of California: California Min. Bur. Bull. 23, 282 pp., (1902) (Amador County, pp. 182-187).

Aubury, Lewis E., 1906, Structural and industrial materials of California: California Min. Bur. Bull. 38, 412 pp.

Aubury, L. E., 1908, The copper resources of California: California Min. Bur. Bull. 50, 366 pp. (Amador County, pp. 221-228).

Averill, C. V., 1946, Placer mining for gold in California: California Div. Mines Bull. 135, pp. 231-233.

Averill, C. V., and Norman, L. A., Jr., eds., 1951, Counties of California, mineral production and significant mining activities of 1949: California Jour. Mines and Geology, vol. 47 (Amador County, pp. 310-312).

Bates, T. F., 1945, Origin of the Edwin clay, Ione, California: Geol. Soc. America Bull., vol. 56, pp. 1-38.

Boalich, E. S., Castello, W. O., Huguenin, E., Logan, C. A., Tucker, W. Burling, 1920, The clay industry in California: California Min. Bur. Prel. Rept. 7, pp. 38-42.

Bowen, O. E., Jr., and Crippen, R. A., Jr., 1948, Geologic maps and notes along highway 49: California Div. Mines Bull. 141, pp. 62-65.

Bradley, W. W., Huguenin, E., et al., 1918 Manganese and chromium in California: California Min. Bur. Bull. 76, pp. 29-30, 116-118.

Browne, J. R., 1868, Mineral resources of the states and territories west of the Rocky Mountains for 1867: U. S. Treasury Dept., p. 80.

Cater, F. W., Jr., 1948, Chromite deposits of Calaveras and Amador Counties: California Div. Mines Bull. 134, pt. III, chap. 2.

Crawford, J. J., 1894, Twelfth report (second biennial) of the State Mineralogist, for two years ending September 15, 1894: California Min. Bur. Rept. 12, 541 pp. (Amador County: gold, pp. 70-80.)

Crawford, J. J., 1896, Thirteenth report (third biennial) of the State Mineralogist, for two years ending September 15, 1896: California Min. Bur. Rept. 13, 726 pp. (Amador County: gold, pp. 65-81.)

Dietrich, W. F., 1928, The clay resources and the ceramic industry of California: California Div. Mines and Mining Bull. 99, pp. 49-63.

Eakle, Arthur S., 1922, Minerals of California: California Min. Bur. Bull. 91, pp. 13-18.

Fairbanks, H. W., 1890, Geology of the Mother Lode Region: California Min. Bur. Rept. 10, pp. 23-90.

Goodyear, W. A., 1877, Coal mines of the western coast of the United States: A. L. Bancroft and Co., p. 79.

Hanks, Henry G., 1886, Sixth report of the State Mineralogist for the year ending June 1, 1886: California Min. Bur. Rept. 6, 2 parts; pt. I, 145 pp.; pt. II, 222 pp. (Amador County, pt. II, pp. 15-24.)

Heyl, George R., 1948, Foothill copper-zinc belt of the Sierra Nevada, California: California Div. Mines Bull. 144, pp. 15-19.

Heyl, George R., and Eric, John H., 1948, Newton copper mine, Amador County, California: California Div. Mines Bull. 144, pp. 49-60.

Irelan, Wm., Jr., 1888, Eighth report of the State Mineralogist for two years ending October 1, 1888: California Min. Bur. Rept. 8, 948 pp. (Amador County, pp. 41-115.)

Irelan, Wm., Jr., 1890, Tenth report of the State Mineralogist for the year ending December 1, 1890: California Min. Bur. Rept. 8, 983 pp. (Amador County, pp. 98-123.)

Irelan, Wm., Jr., 1892, Eleventh report of the State Mineralogist for two years ending September 15, 1892: California Min. Bur. Rept. 11, 612 pp. (Amador County, pp. 139-149.)

Jenkins, Olaf P., 1938, Geologic map of California: California Div. Mines, scale 1 : 500,000, 6 sheets.

Jenkins, Olaf P., ed., 1948a, Copper in California: California Div. Mines Bull. 144, 429 pp. (Amador County, pp. 49-60, 85-91, 214-216, pl. 7-13, 33.)

Jenkins, Olaf P., 1948b, Outline geologic map of California showing locations of copper properties: California Div. Mines economic mineral map no. 6.

Jenkins, Olaf P., ed., 1949, Counties of California, mineral resources and mineral production during 1947: California Div. Mines Bull. 142, p. 35. (Amador County, pp. 33-35.)

Jenkins, Olaf P., ed., 1950, Mineral commodities of California: California Div. Mines Bull. 156, 443 pp.

Johnson, F. T., and Ricker, Spangler, 1948, Ione-Carbondale clays, Amador County, California: U. S. Bur. Mines Rept. Inv. 4213, 6 pp.

Johnson, F. T., and Ricker, Spangler, 1949, Ione-Carbondale clays, Amador County, California: California Div. Mines, Rept. 45, pp. 491-498.

Knopf, Adolph, 1929, The Mother Lode system of California: U. S. Geol. Surv. Prof. Paper 157, 88 pp.

Kunz, George F., 1905, Gems, jewelers' materials and ornamental stones of California: California Min. Bur. Bull. 37, pp. 40-41.

Lang, Herbert, 1907, The copper belt of California: Eng. Min. Jour., vol. 84, pp. 909-913.

Lindgren, W., and Turner, H. W., 1894, U. S. Geol. Survey Geol. Atlas, Placerville folio (no. 3), 9 pp., 4 maps.

Lindgren, Waldemar, 1895, Characteristic features of California gold-quartz veins: Geol. Soc. America Bull., vol. 6, pp. 221-240.

Lindgren, Waldemar, 1896, U. S. Geol. Survey Geol. Atlas, Pyramid Peak folio (no. 31), 8 pp., 4 maps.

Lindgren, Waldemar, 1911, The Tertiary gravels of the Sierra Nevada of California: U. S. Geol. Survey Prof. Paper 73, pp. 21-28, 46-48, 196-197, pl. 1.

Lindgren, Waldemar, 1928, Mineral deposits, 3d ed. 1049 pp., New York, McGraw-Hill Book Co.

Logan, C. A., 1921, Auburn field division—Amador County: California Min. Bur. Rept. 17, pp. 412-413.

Logan, C. A., 1923a, Auburn field division: California Min. Bur. Rept. 18, p. 7.

Logan, C. A., 1923b, Auburn field division—Amador County: California Min. Bur. Rept. 19, pp. 94-96, 143.

Logan, C. A., 1924, Sacramento field division: California Min. Bur. Rept. 20, pp. 73-84 (Amador County, pp. 73-75).

Logan, C. A., 1927, Amador County: California Min. Bur. Rept. 23, pp. 131-202.

Logan, C. A., 1934, Mother Lode gold belt of California: California Div. Mines Bull. 108, 240 pp. (Amador County, pp. 55-123.)

Logan, C. A., 1947, Limestone in California: California Div. Mines Rept. 43, pp. 175-357.

Murdoch, Joseph, and Webb, R. W., 1948, Minerals of California: California Div. Mines Bull. 136, p. 37.

Norman, L. A., Jr., 1939, Operations at the Old Eureka mine: Mining Technology, Am. Inst. Min. Met. Eng., Technical Publication No. 1136, vol. 3, no. 6.

Pabst, Adolph, 1938, Minerals of California: California Div. Mines Bull. 113, 344 pp. (Amador County chalcedony, pp. 96, 99.)

Pask, Joseph A., and Turner, Mort D., 1952, Geology and ceramic properties of the Ione formation, Buena Vista area, Amador County, California: California Div. Mines Special Rept. 19, 39 pp.

Piper, A. M., Gale, H. S., Thomas, H. E., and Robinson, T. W., 1939, Geology and ground water hydrology of the Mokelumne area, California: U. S. Geol. Survey Water-Supply Paper 780, 230 pp., 22 plates.

Raymond, Rossiter, W., 1874, Statistics of mines and mining in the states and territories west of the Rocky Mountains, being the fifth annual report of Rossiter W. Raymond, United States Commissioner of Mining Statistics, Washington, D. C., Government Printing Office, 585 pp.

Sampson, R. J., and Tucker, W. B., 1931, Feldspar, silica, andalusite, and cyanite deposits of California: California Div. Mines Rept. 27, pp. 432-435.

Sawyer, F. G., 1949, Montan by the mountain: Ind. and Eng. Chemistry, vol. 41, no. 4, pp. 14a-16a, April.

Schuette, C. N., 1929, Cutting clay production costs with systematic mining methods: Eng. and Min. Jour. vol. 127, pp. 196-198.

Sperisen, Francis J., 1938, Gem minerals of California: California Div. Mines Rept. 34, pp. 34-78.

Storms, W. H., 1900, Mother Lode region of California: California Min. Bur. Bull. 18, 154 pp.

Taliaferro, N. L., 1943, Manganese deposits of the Sierra Nevada, their genesis and metamorphism: California Div. Mines Bull. 125, pp. 277-332.

Trask, P. D., Wilson, I. F., and Simons, F. S., 1943, Manganese deposits of California, a summary report; Manganese in California: California Div. Mines Bull. 125, pp. 51-215 (Amador County, p. 75.)

Tregloan, J. B., 1903, Register of mines and minerals, Amador County, California: California Min. Bur. Register of Mines, no. 12, pp. 13-14.

Tucker, W. B., 1914, Amador County: California Min. Bur. Rept. 14, pp. 3-53.

Turner, H. W., 1894, U. S. Geol. Survey Geol. Atlas, Jackson folio (no. 11), 8 pp., 4 maps.

Turner, H. W., and Ransome, F. L., 1898, U. S. Geol. Survey Geol. Atlas, Big Trees folio (no. 51), 8 pp., 1 sheet illus., 3 maps.

Turner, Mort D., 1950a, Sand and gravel: California Div. Mines Bull. 156, p. 257.

Turner, Mort D., 1950b, Slate: California Div. Mines Bull. 156, p. 259, 1950.

U. S. Bureau of Mines, 1943, West belt copper-zinc mines, Amador and Calaveras Counties, California: U. S. Bur. Mines War Minerals Rept. 103, 16 pp., 1943.

Watts, W. L., 1892, Coal and clays in Amador County: California Min. Bur. Rept. 11, pp. 148-149.

Whitney, J. D., 1865, Geology of California, vol. 1, p. 269.

Wright, Lauren A., 1950, Talc, soapstone, and pyrophyllite: California Div. Mines Bull. 156, p. 277.

TABULATED LIST OF MINES AND PROSPECTS IN AMADOR COUNTY

The following list of mines and prospects in Amador County is arranged alphabetically by commodity and by name of prospect or mine. Numbers which appear in the left-hand column under the heading *Map No.* refer to the *Geologic Map of Amador County Showing Mines and Mineral Deposits,* Plate 1.

CHROMITE

MAP NO.	CLAIM, MINE, OR GROUP	OWNER NAME, ADDRESS	LOCATION SEC.	T.	R.	B & M	REMARKS
1	Allen property (Johnson lease)	George Allen, Sutter Creek	27, 28	6N	10E	MD	Two shafts, 10 feet apart, 12 and 30 feet deep and shallow cut. An 18 to 20 inch chromite lens exposed bottom of 12-foot shaft worked out? (Cater 48:55)
2	Carr and Mefford (Mooney claim)	Frank and Rose Devencenzi, Volcano	W½ SE¼ 34	6N	10E	MD	Chromite mined from open cuts and some underground workings which are now largely caved. Last worked 1918. (Bradley et al. 18:116; Cater 48:53)
	Courtwright property (Daggett lease)	Carl H. Kremmel, R.F.D. Ione	SE¼ 3	5N	10E	MD	Chromite mined by adits, raises, and shafts which are flooded and inaccessible. Last worked 1916. (Bradley et al. 18:117; Cater 48:53)
	Courtwright property (Woods and Roach lease)	Emma Garibaldi, Jackson	NE¼ SW¼ 34	6N	10E	MD	Chromite mined from two shafts, 30 and 32 feet deep. Ore occurs as small lenses in a schistose serpentine (Cater 48:54)
	Detert property	W.F. Detert Estate 1715 Mills Tower, San Francisco	near center 6	7N	10E	MD	Workings all shallow. Much of chromite ore was plowed from soil overlying serpentine. Last worked World War I. (Bradley et al. 18:117; Cater 48:55-56)
3	Ellis property	J.M. Ellis, Box 98 Jackson	NW¼ SW¼ 2	5N	10E	MD	Workings consist of one small pit and a number of prospect trenches. Lenses of chromite in serpentine are small, ranging from 1 to 3 feet thick and up to 25 feet long. Last worked World War I. (Cater 48:53)
	Fields	Adrian Fredricks, Ione	SW¼ 10	5N	10E	MD	Chromite mined from inclined shaft. The ore occurs as lenses with some uvarovite. Last worked 1918. (Cater 48:52)
	Kremmel and Froelich	Fred W. DuFrene et al. Ione	NW¼ NE¼ 34	6N	10E	MD	Chromite mined from small open cut and adit with stoping to surface. Small stringers of disseminated and massive ore in serpentine. Last worked 1916. (Cater 48:54)

CHROMITE, (cont.)

| MAP NO. | CLAIM, MINE, OR GROUP | OWNER NAME, ADDRESS | LOCATION | | | | REMARKS |
			SEC.	T.	R.	B & M	
	Taylor Ranch	W.F. Detert Estate 1715 Mills Tower, San Francisco	NE¼ 6	7N	10E	MD	Chromite mined from shallow pits. Irregular lenses of ore in highly sheared serpentine (Cater 48:56)
	Wait property	A.L. and R.A. Wait Plymouth	NW¼ 29	7N	10E	MD	Several small pods and lenses chromite mined from open cuts. Last worked 1916. (Bradley et al. 18:117; Cater 48:55)

COPPER

MAP NO.	CLAIM, MINE, OR GROUP	OWNER NAME, ADDRESS	SEC.	T.	R.	B & M	REMARKS
	Allen: See Hayward						
	Bull Run	East Bay Municipal Utility District Oakland	15	5N	10E	MD	Shipped during sixties; 400-foot shaft. (Aubury 05:186; 08:225; Logan 27:149; Jenkins 48:214)
	Chaparral	East Bay Municipal Utility District Oakland	10	5N	10E	MD	Opened 1864; 120-foot shaft (Aubury 05:186; 08:225; Jenkins 48:214)
4	Copper Hill	W.F. Detert Estate, 1715 Mills Tower, San Francisco	34 35	8N	9E	MD	(Aubury 05:186-187; 08:226-227; Logan 27:148; Raymond 74:87; Tucker 14:12-13; Jenkins 48:214; herein)
	Forest Home	Annie L. Devore, Plymouth	9	7N	9E	MD	Four shafts 80 feet deep. (Aubury 05:187; 08:227; Jenkins 48:215)
	Hayward (Allen)	Allen Estate Co., Sutter Creek	28	8N	10E	MD	Chalcopyrite associated with pyrite and some sphalerite (Tucker 14:12; Jenkins 48:215, pl. 13)
	Ione City	Pearl T. Fye, Ione	3, 4	5N	10E	MD	Operated in sixties (Aubury 05:186; 08:225; Jenkins 48:215)
	Ione Copper (Irish Hill)	Crocker First Nat'l. Bank, San Francisco	2	6N (Arroyo Seco)	9E	MD (Arroyo Seco) proj.	In greenstone just east of slate belt. Reported to have produced some stibnite as well as copper ores. On Arroyo Seco grant. (Aubury 08:227; Lang 07:909, 964, 966; Logan 27:149; Tucker 14:13; Jenkins 48:215)
	Irish Hill: See Ione Copper						
	Johnson Ranch	Frank Boscovich, Jackson	25,35 36	6N	10E	MD	Shafts 90, 60 feet deep; vein up to 12 feet. (Aubury 08:227; Tucker 14:13; Jenkins 48:215)

COPPER, (cont.)

MAP NO.	CLAIM, MINE, OR GROUP	OWNER NAME, ADDRESS	LOCATION				REMARKS
			SEC.	T.	R.	B. & M.	
	Moon	Lorraine McCarthy Davis, 747 La Paloma Road, Richmond	3,9, 10	5N	10E	MD	Shafts 140 and 100 feet deep. (Aubury 05:185-186; 08:224-225; Tucker 14:13; Jenkins 48:216)
5	Newton	Fred W. DuFrene et al, Ione	28	6N	10E	MD	(Aubury 05:183-185; 08:222-224; Browne 67:143, 149, 166-167; Irelan 88:106-108; Logan 27:148; Raymond 74:47-87; Tucker 14:13; Jenkins 48:216, 49-60, pl. 7-11; herein)
	Russel	Adrian Fredricks, Ione	10	5N	10E	MD	Large dump; 200-foot shaft (Aubury 05:186; 08:225; Jenkins 48:216)
	Thayer (Grayhouse, Grey House)	East Bay Municipal Utility District, Oakland	23	5N	10E	MD	Shaft 240 feet deep; drilled by U.S. Bureau of Mines 1943. (Aubury 05:186; 08:225; Tucker 14:14; Jenkins 48:85-91, 216, pl. 33)

GOLD, LODE

MAP NO.	CLAIM, MINE, OR GROUP	OWNER NAME, ADDRESS	SEC.	T.	R.	B & M	REMARKS
	A and B	Reaves, M.H. & L.D. c/o L.D. Reaves, 1704 N St., Sacramento 14	11,14	7N	10E	MD	Inclined shaft 175-ft. deep, two crosscut adits, and 200-ft. open cut. (Tucker 14:14; Logan 27:186; Logan 35:80)
	Acme: See Fort Ann						(Crawford 96:66, Tucker 14:14; Logan 27:186)
	Acme Cons. et al.	E.J. and J.M. Muldoon 206 Auburn St., Modesto	9	5N	11E	MD	(Logan 27:186)
	Aetna: See Amador Gold						
6	Alma	Ralph McGee, Sutter Creek	29	6N	11E	MD	Main shaft is 1000 ft. deep. Some work done in 1948 in smaller 115-ft. deep shaft. (Crawford 94:70; Crawford 96:66-67; Tucker 14:14-15; Logan 27:186)
	Alphi	Lauren E. Wilkenson 617 W. Pine St., Lodi	34,35	6N	11E	MD	(Logan 34:117)
	Alta	L.F. and Lillie Payton, Pine Grove	28	7N	12E	MD	
	Alta extension	L.F. and Lillie Payton, Pine Grove	28	7N	12E	MD	
7	Amador Gold	Ralph McGee, Sutter Creek	34	6N	11E	MD	Most work done before 1900. Mine has 800-ft. inclined shaft and five levels (Irelan 88:88-81; Crawford 94:71; 96:72; Tucker 14:15; Storms 00:44; Logan 27:186; Logan 34:60,117)
	Amador Queen #1: See Amador Gold	Ralph McGee, Sutter Creek	34,35	6N	11E	MD	

GOLD, LODE, (cont.)

MAP NO.	CLAIM, MINE, OR GROUP	OWNER NAME, ADDRESS	LOCATION				REMARKS
			SEC.	T.	R.	B & M	
8	Amador Queen #2	Ralph McGee, Sutter Creek	34,35	6N	11E	MD	(Irelan 88:9,93; Irelan 90:107; Crawford 94:71; Storms 00:44; Tucker 14:16; Logan 27:186; Knopf 29:70; Logan 34:60,61,117; herein)
9	Amador Star	Kaiser, Eliza M., 720 West Poplar St., Stockton	23	8N	10E	MD	(Logan 27:186; Logan 34:61-62; Jenkins 48:65-66; herein)
	Anita	O'Grady, Geo. E., Jackson	28	6N	11E	MD	750-ft. inclined shaft (Crawford 96:68; Tucker 14:16; Logan 34:117)
10	Argonaut	B. Monte Verda and E.C. Taylor, 369 Pine St., San Francisco	17,20,21	6N	11E	MD	(Crawford 96:68; Tucker 14:17-19; Logan 21:406-408; 22:23-24; 23:13-14; 24a:2; 24c:177; 27:153-157,186; 34:62-70,117; Storms 00:50-52; Knopf 29:66-68; Jenkins 48:64; herein)
	Atlantic	Detert, W.F. Est., 1715 Mills Tower Building San Francisco	13,14	7N	10E	MD	(Logan 34:117)
11	Ballard	Ballard Mother Lode Mines, Ltd. c/o John Ratto, Sutter Creek	14 23	8N	10E	MD	(Logan 34:70,117; herein)
	Baratini	East Bay Municipal Utility District, Oakland	15	5N	11E	MD	
	Battle Mountain	William Logomarsino et al, Volcano	22	7N	12E	MD	
12	Bay State	California Lands Inc. c/o Capital Co., #1 Powell St., San Francisco	23	8N	10E	MD	Opened in early 1890's. Main shaft is 1065 ft. inclined with 7 levels. Last production was 1909. Reopened in 1933 but nothing produced. Shaft dewatered in late 1934. Work done on 300 level in early 1935. (Irelan 93:146; Crawford

GOLD, LODE, (cont.)

MAP NO.	CLAIM, MINE, OR GROUP	OWNER NAME, ADDRESS	LOCATION SEC.	T.	R.	B & M	REMARKS
12	Bay State (cont'd)						94:71; 96:69; Tucker 14:19-20; Storms 00:85-86; Logan 27:187; 34:71,117; Jenkins 48:65-66)
13	Belden	Belden Amador Mines Inc., P.O. Box 28, Fort Wayne, Indiana	26	7N	13E	MD	Herein
	Bellwether	Mark Eudey, Martell	21	6N	11E	MD	Mine development began 1895. Several shallow shafts in addition to a 180-ft. and a 320-ft. shaft. (Irelan 90:104-105; Irelan 93:141; Crawford 94:71; 96:69; Logan 27:187; 34:71-72, 117)
	Belmont	Fred Suelberger c/o Esther J. Fitzgerald 252 Los Flores, Bakersfield	32	7N	13E	MD	Gold-quartz vein strikes north, dips 80° W in granodiorite. Some work done on 100-ft. level off main 145-ft. shaft and another shaft was sunk during 1929. (Tucker 14:20)
	Black Hills:See Italian						
	Black Metal: See Valpariso						
14	Black Prince	W.R. Schwickerth and P.M. Wedell, c/o W.R. Schwickerth, Pine Grove	27	7N	13E	MD	Herein
15	Black Wonder	George B. and Florence J. Taves, Pine Grove	5	6N	12E	MD	Herein
	Boro	Mary Gilbert, 2317 24th Ave., San Francisco 16	24	7N	10E	MD	(Logan 34:117)

GOLD, LODE, (cont.)

MAP NO.	CLAIM, MINE, OR GROUP	OWNER NAME, ADDRESS	LOCATION				REMARKS
			SEC.	T.	R.	B & M	
	Bona Fortuna	Glen and Everett Fancher, Jackson	25	7N	10E	MD	(Logan 34:117)
	Bright: See Anita						
	Brown: See Jose Gulch						
	Bruce	Bruce Cornwall, 57 Sutter St., San Francisco	14	7N	10E	MD	(Logan 34:117)
16	Bunker Hill	Bunker Hill Mining Co. c/o C.E. Crandall North Bend, Washington	25 36	7N	10E	MD	(Irelan 88:57-63; 90:75,114; 93:79; Crawford 94:79; 96:77; Storms 00:80; Tucker 14:20-21; Logan 21:408-409; 22:24; 23:298-299; 23:2,15; 24a:2; 27:157-158,187; 34:72-74; Knopf 29:55-56; Jenkins 48:65; herein.)
	California	W.T. Kneen, Box 351, Mooreland, Oklahoma	20	7N	13E	MD	See Lone Willow group.
	California group	W.F. Detert Estate 1715 Mills Tower, San Francisco	24	7N	10E	MD	First worked in 1852. Some work done during middle 1930's. (Logan 27:187; 34:81.)
	Caucasian Consolidated	Ralph McGee, Jackson	23	8N	10E	MD	Along 600 ft. of contact between greenstone and slate. quartz crops out on the surface for short distances. Three shafts 104,85, and 85 ft. deep and a 500-ft. adit are located on this property (Irelan 90:120; Logan 27:188; 34:118.)
17	Central Eureka (including Old Eureka)	Central Eureka Mining Co., Russ Bldg., San Francisco	7,8	6N	11E	MD	(Irelan 88:49; 90:104,72,102,113; 93:144; Crawford 94:79; 96:67,69-70; Hamilton 14:24,28; Logan 21:409-410,411; 22: 24; 23:299; 23:215,73; 24a:2,3; 24b:74; 27:158-164,176-178,188,194; 34:74-80,101,104,118,121; Storms 00:64-65;

GOLD, LODE, (cont.)

MAP NO.	CLAIM, MINE, OR GROUP	OWNER NAME, ADDRESS	SEC.	T.	R.	B & M	REMARKS
17	Central Eureka (including Old Eureka) (continued)						Jenkins 48:64; herein.)
	Chief	J.J. Young, 32 Bellevue Ave., Piedmont	14	7N	10E		(Logan 27:188.)
	Chili Jim	2/5 Bruce Cornwall 57 Sutter St., San Francisco, 3/5 W.F. Detert Est., 1715 Mills Tower, San Francisco	24	7	10		(Logan 27:188; 34:118.)
	Clyde	Mark Eudey, Martell	17 20	6N	11E	MD	Part of Kennedy mine (Irelan 90:110-111; Logan 27:188; 34:118.)
	Comet	Baumhardt, Mary M. et al c/o Helen Agen, Rt. 5, Box 2060, Modesto	6	6N	11E	MD	Shaft is 200 ft. deep. (Irelan 90:114; Tucker 14:26; Logan 27:188; 34:118.)
18	Contini	Bert Contini, Jackson	9	6N	12E	MD	Herein.
	Cons. McNamara	George Newman, 2047 33rd Ave., San Francisco	3	5N	11E	MD	(Logan 27:188; 34:118.)
	Cosmopolitan	W.F. Detert Est., 1715 Mills Tower, San Francisco	14	7N	10E	MD	Worked between 1850 and 1890. Shaft was sunk 750 ft. with 6 levels (Logan 27:188; 34:118.)
	Coulter	Hobart Estate Co., and Emma Rose, 202 Hobart Bldg., San Francisco	14	7N	10E	MD	

GOLD, LODE, (cont.)

MAP NO.	CLAIM, MINE, OR GROUP	OWNER NAME, ADDRESS	SEC.	T.	R.	B & M	REMARKS
	Creek Ledge	Tony Levaggi, et al., Plymouth	35	8N	10E	MD	Mined through a 100-ft. inclined shaft. (Tucker 14:27; Logan 27:188; 34:80,118.)
	Croesus and St. Martin	Levaggi, Elvira, c/o James B. Levaggi, 2752 Buchanan St., San Francisco	26	8N	10E	MD	(Logan 34:118.)
19	Crown Point	John M. and Emma Fontenrose, Jackson	22 23	7N	10E	MD	Deepest work done is at Bonanza shaft which is 485 ft. (Irelan 90:116; Logan 27:118.)
20	Defender	West Point Consolidated Mines Co., c/o E.H. Outerbridge, 250 Park Ave., New York, New York	29	7N	13E	MD	(Tucker 14:27; Hamilton 22:24; Logan 24b:75; herein.)
	Defiance	West Point Consolidated Mines Co., c/o E.H. Outerbridge, 250 Park Ave., New York, New York	29	7N	13E	MD	
	Dewitt	Ralph McGee, Sutter Creek	3	5N	11E	MD	Some work done in 1933. (Logan 34:118.)
	Douglas	C.M. Farnsworth, Carrie Mason	28	7N	12E	MD	
	Dowling		27 28	6N	11E	MD	(Logan 34:118.)
	Downs	Peter Barone, Volcano	13	7N	12E	MD	Originally located in 1877. (Irelan 87:20; Crawford 94: 72; 96:70; Tucker 14:27; Logan 27:189.)

GOLD, LODE, (cont.)

MAP NO.	CLAIM, MINE, OR GROUP	OWNER NAME, ADDRESS	SEC.	T.	R.	B & M	REMARKS
	Doyle	Ralph McGee, Sutter Creek	34	6N	11E	MD	Shaft 150 ft. deep. (Irelan 90:107; Crawford 94:72; 96:70; Tucker 14:27; Logan 27:189; Logan 34:118.)
	Dry Creek						See Cosmopolitan.
	Drytown Consolidated						See Crown Point.
	East Amador	J.W. Bullock, c/o C.E. Crandall, North Bend, Washington	36	7N	10E	MD	
	Ebelbrau	R.L., Thelma L. Wolfe Fiddletown	23	8N	11E	MD	(Logan 27:189.)
	Eclipse						Part of Original Amador.
	El Dorado	T.C. Mayon et al, c/o Mrs. R.W. Emerson, Box 85, Centerville, Calif.	36, 31	7N, 7N	10E, 11E	MD, MD	Shaft 300 ft. deep. (Irelan 90:99; Tucker 14:27; Logan 27:189; 34:82,118.)
	Elephantine	Daisy Thomas, Jackson	10	5N	11E	MD	(Logan 27:189.)
21	Elkhorn	Mary and Sadie R. Grillo, Volcano	32	7N	13E	MD	Herein.
	Empire	Ralph McGee, Sutter Creek	3	5N	11E	MD	(Logan 35:118.)
	Eureka	Pacific Gas & Electric Co., San Francisco	27, 28	7N	13E	MD	See Old Eureka under Central Eureka. (Irelan 90:72; Crawford 96:67; Tucker 14:28; Logan 27:189; 35:118.)
	Eureka #2	W.F. Detert estate 1715 Mills Tower, San Francisco	23	7N	10E	MD	(Logan 34:119.)

GOLD, LODE, (cont.)

MAP NO.	CLAIM, MINE, OR GROUP	OWNER NAME, ADDRESS	LOCATION SEC.	T.	R.	B & M	REMARKS
	Excelsior	Gertrude Clark, Sutter Creek	23	8N	10E	MD	(Logan 27:189; 34:119.)
	Excelsior	J.W. Bullock, c/o C.E. Crandall, North Bend Washington	30	7N	11E	MD	Shaft is 50 feet deep (Tucker 14:28; Logan 27:189; 34:119.)
	Farrell	Joe Rosenberg, c/o Harvey N. Seligman, 1805 Sharp Ave., Walnut Creek	15	5N	11E	MD	Some activity in 1937. (Irelan 90:107-108; Crawford 94:72; 96:70; Tucker 14:28; Logan 27:189; 34:119.)
	Florence	Kenneth W. Dennison 636 33rd Ave., San Francisco 21	4	6N	12E	MD	(Logan 27:189.)
22	Fort Ann	William L. Metcalf Box 32, Volcano	W½ 2	7N	12E	MD	Formerly known as Acme and Robinson. (Crawford 96:66; Tucker 14:14; Logan 27:186; herein.)
	Fort John	Emma Rose, c/o E.C. McCurdy, Mills Bldg. 220 Montgomery St., San Francisco	11	7N	10E	MD	(Logan 27:189; 34:119.)
	Forty-nine	E.S. Barney Est., c/o Anna M. Irving, 1505 Hillcrest Ave., Orlando, Florida	14	7N	10E	MD	(Irelan 90:122; Logan 27:189; 34:119.)
23	Fremont-Gover	Fremont Gover Co., c/o C.B. Braun, Sec'y., Rt. 3, Box 306, Albany Oregon	25	7N	10E	MD	(Irelan 88:53-57; 90:75; 93:146; Crawford 94:72-74; 96:71 Storms 00:80-81; Tucker 14:28-29; Logan 21:411; 22:24; 23:299; 23:15,73,143; 24a:2; 27:165-166,190; Knopf 29:52-54; Logan 34:82-84,119; Jenkins 48:65; herein.)

GOLD, LODE, (cont.)

MAP NO.	CLAIM, MINE, OR GROUP	OWNER NAME, ADDRESS	SEC.	T.	R.	B & M	REMARKS
	Goat Ranch	W.F. Detert Estate 1715 Mills Tower, San Francisco	14	8N	10E	MD	
	Golden Crown	Wendell A. Boitano Sutter Creek	18,19	7N	11E	MD	(Logan 27:190.)
	Golden Eagle	Helen Crotty, Sutter Creek	6	6N	11E	MD	(Irelan 90:113-114; Logan 27:190; 34:119.)
	Golden Gate	John F. Davis Estate 2872 Jackson St., San Francisco	14	7N	10E	MD	(Logan 27:190; 34:119.)
	Golden Gate	W.F. Detert Estate 1715 Mills Tower, 220 Bush St., San Francisco and Geo. Kretcher Jr. Plymouth	14	8N	10E	MD	(Logan 27:190; 34:119.)
	Golden Gate	Chas. W. Cassinelli c/o A. Marre, 643 Chetwood St., Oakland	12 13	7N	12E	MD	
24	Gold Mountain	Gold Mountain Mining Co., c/o Wm. D. Shea Jr., 1249 Russ Bldg. San Francisco	19	7N	11E	MD	Active in 1890's, shaft sunk in 1935, some prospecting done in 1945. Extensive quartz outcroppings, free gold with pyrite and small amounts arsenopyrite and galena. Silver content high. Adit and 100-ft shaft. (Irelan 88: 94; 93:145; Crawford 94:72; 96:71; Logan 27:190.)
	Gover						See Fremont.
	Governor Bradford	W.F. Detert Estate 1715 Mills Tower, San Francisco	14	7N	10E	MD	

GOLD, LODE, (cont.)

| MAP NO. | CLAIM, MINE, OR GROUP | OWNER NAME, ADDRESS | LOCATION | | | | REMARKS |
			SEC.	T.	R.	B & M	
	Great Eastern	J.W. Bullock, c/o C.E. Crandall, North Bend Washington	36	7N	10E	MD	
	Great Eastern	Great Eastern Mining Co., c/o Elmer Tripp Plymouth	23	8N	10E	MD	(Logan 27:190; 34:119.)
	Green	J.J. Young, 32 Belle-vue, Piedmont	14	7N	10E	MD	
	Grey Eagle	O. Ball, c/o Laura C. Bell, Plymouth	2	7N	10E	MD	
25	Hageman	Frank E. Cotie and Vernie Hoffschneider	N½33	7N	13E	MD	(Logan 23:7; 27:190; herein.)
26	Hardenburg	Ralph McGee, Sutter Creek	10 3	5N 6N	11E 11E	MD MD	Vertical shaft 1500 ft. deep. Last worked in 1918 (Irelan 90:68,106-107; 93:139-140; Crawford 94:74; 96:71; Tucker 14:29-30; Logan 27:166,190; 34:84-85,119.)
	Hayvire						See Elkhorn.
	Hazard						See Treasure.
	Henry Clay						See Cosmopolitan.
	Hercules	W.F. Detert Estate 1715 Mills Tower, San Francisco	14	7N	10E	MD	(Logan 27:191; 34:119.)
	Homestake	Alex C. and Mae M. Matulich, Drytown	14	7N	10E	MD	
	Honula	West Consolidated Mines Co., c/o E.H. Outerbridge, 250 Park Ave.	32	7N	13E	MD	

GOLD, LODE, (cont.)

MAP NO.	CLAIM, MINE, OR GROUP	OWNER NAME, ADDRESS	SEC.	T.	R.	B & M	REMARKS
	Honula (cont'd)	New York, New York					
	Illinois, Green, et al.	J.J. Young, 32 Bellevue Ave., Piedmont	35	8N	10E	MD	(Irelan 90:119; Logan 27:191; 34:119.)
	Indiana	B. Monte Verda and E.C. Taylor, 369 Pine St. San Francisco	11	7N	10E	MD	Part of Plymouth Consolidated.
	Isaac Newton	James C. Harle, c/o Milbank, Tweed, Hope et al., 15 Broad St. New York, New York	31	6N	11E	MD	Adit 300 ft. long (Tucker 14:30; Logan 27:191.)
27	Italian	Black Hills Mining Co. c/o Wm. Tam, Jackson	24	7N	10E	MD	(Irelan 90:115-116, Logan 27:191; 34:85,119; Jenkins 48:65; herein.)
	Jack	West Point Consolidated Mines Co., c/o E.H. Outerbridge, 250 Park Ave., New York, New York	32	7N	13E	MD	
	Jackson	B. Monte Verda and E.C. Taylor, 369 Pine St. San Francisco	20	6N	11E	MD	(Logan 35:119.)
	Jill	West Point Consolidated Mines Co., c/o E.H. Outerbridge, 250 Park Ave., New York, New York	32	7N	13E	MD	
	Joe Davis	W.F. Detert Estate	13,14	7N	10E	MD	(Logan 34:119.)

GOLD, LODE, (cont.)

MAP NO.	CLAIM, MINE, OR GROUP	OWNER NAME, ADDRESS	SEC.	T.	R.	B & M	REMARKS
	Joe Davis (cont'd)	1715 Mills Tower, San Francisco					
	Jose Gulch (Brown Aurora et al.)	Joseph A. Lema, Jackson	2 35	5N 6N	11E 11E	MD MD	200-ft. shaft. (Tucker 14:30; Logan 27:191; 34:117.)
	Julia	M.C. Quinn, c/o Mrs. E. Sargent, Jackson	10	5N	11E	MD	
28	Jumbo	Behrend Doscher, Pioneer Station	SW¼ 26	7N	13E	MD	Herein.
	Jupiter	M.F. Gates, c/o Alex R. Smith, 700 San Antonio Way, Sacramento, and Alex R. Smith	21 22	8N	11E	MD	(Logan 27:191.)
	Kelly	Hoit C. Vicini and Margaret V. Hobbs, 48 Maple Ave., Atherton	3	5N	11E	MD	Shaft 90 ft. deep. (Irelan 90:106; Crawford 94:74; 96:72; Tucker 14:30; Logan 27:191; 35:120.)
29	Kennedy	Mark and Frances Eudey, Martell	16,17 20,21	6N	11E	MD	(Irelan 88:66-71; 90:103-104; 93:141-142; Crawford 94:77; 96:72; Storms 00:52,60; Tucker 14:31-34; Logan 21:408; 23:299; 23:15; Root 24a:2-3,177; 27:166-169,191; Knopf 29:63-66; Logan 34:85-91,120; Jenkins 48:62,64; herein.)
	Kennedy Extension	Mark and Frances Eudey, Martell	20	6 N	11 E	MD	(Logan 27:191.)
30	Keystone	Keystone Mining Co. c/o John D. Culbert Amador City	36	7N	10E	MD	(Hanks 86: pt. II, 16-17; Irelan 88:63-66; 90:72-73,98; 93:145; Crawford 94:77; 96:72; Storms 00:77-80; Tucker 14: 34-36; Logan 21:411; 27:169-170; Knopf 29:58-59; Logan 34: 91-95; Jenkins 48:65; herein.)

GOLD, LODE, (cont.)

MAP NO.	CLAIM, MINE, OR GROUP	OWNER NAME, ADDRESS	SEC.	T.	R.	B & M	REMARKS
	Kruger & Vaughn	W.W. Steele, Jackson 1/10 int., Geo. M. Thomas, Jackson, 9/10 interest	3	5 N	11E	MD	Greenstone hanging wall, slate footwall. (Irelan 90:107; Logan 27:192; 34:95.)
	Last Chance	J.W. Bullock, c/o C.E. Crandall, North Bend Washington	36	7N	10E	MD	240-ft. shaft, wholly in slates. (Irelan 90:115; Logan 27:192; 34:120.)
31	Lincoln						See Lincoln Consolidated.
	Lincoln Consolidated	Lincoln Gold Mining Co., 807 Lonsdale Bldg. Duluth, Minnesota	6,7 8	6N	11E	MD	(Irelan 88:73,75-79; 90:100,101-102; 93:144; Crawford 94: 79-80; 96:73-74,78-79; Storms 00:65-75; Tucker 14:36-37; Logan 27:170-172; 34:95-97; Jenkins 48:64; herein.)
	Little Amador						See Original Amador.
	Littlefield	Ralph McGee, Sutter Creek	10,15	5N	11E	MD	Active in 1893. (Crawford 94:78; 96:74-75; Tucker 14:39; Logan 27:193; 34:121.)
	Little Sargent	Anthony B. Caminetti Jackson	10	5N	11E	MD	(Logan 27:192; 34:120.)
	Lone Willow Group	Joe Porter, Jackson	20	7N	13E	MD	350-ft. tunnel, 100-ft. shaft, tunnel cleaned out and extended in 1935. (Tucker 14:37; Logan 27:192.)
	Lucky Girl	Joseph G. and Mary Dunleavy, Box 28, Pioneer Station	29	7N	13E	MD	
	Lucky Strike, et al.						See Pioneer Lucky Strike.
	Madrona Group						See Pioneer Lucky Strike.
32	Mahoney						See Lincoln Consolidated.

GOLD, LODE, (cont.)

MAP NO.	CLAIM, MINE, OR GROUP	OWNER NAME, ADDRESS	SEC.	T.	R.	B & M	REMARKS
	Mammoth	East Bay Municipal Utility District, Oakland	10	5N	11E	MD	
	Mammoth	Mammoth Mining Co., c/o Anna M. Kerby, Elk Grove	3	6N	12E	MD	
33	Marklee	Marklee Mining Co. c/o T.C. Malloy, 1036 Main St., Napa	1	7N	12E	MD	Idle since 1926. 500-ft. shaft, 800-ft. drift, 1500 feet underground workings. Northwest ore shoot. (Crawford 96:74; Tucker 14:38; Logan 24a: 3; 24b:73-74; 27:172-173.)
	Marlette	Mrs. J.E. Sargent, Jackson	10	5N	11E	MD	See Sargent.
	Maryland	Bruce Cornwall, 57 Sutter St., San Francisco	24	7N	10E	MD	(Logan 34:120.)
	Mayella	W.A. Wilds et al., c/o Mayella Casinelli, Ione	11	7N	10E	MD	(Logan 27:193; 34:120.)
	Mayflower						See Bunker Hill.
	McIntire	J.A. McIntire, c/o H.S. McIntire, 2723 U St., Sacramento	6	6N	11E	MD	210-ft. inclined shaft, 80-ft. shaft, slate hanging wall, greenstone footwall, active in 1895. (Irelan 90:115; Crawford 96:74; Tucker 14:38; Logan 27:193; 34:120.)
	McKay and Love	Hoit C. Vicini and Margaret V. Hubbs, 48 Maple Ave., Atherton	3	5N	11E	MD	See Amador Gold.
	McKinney and Crannis	Mrs. Elsie A. McKinney 506 Park Blvd., Ukiah	10 15	5N	11E	MD	5-ft. vein on slate-greenstone contact, abundant arsenopyrite, 200-ft. adit reopened in 1934. (Irelan 90:106; Logan 27:193; 34:118.)

GOLD, LODE, (cont.)

MAP NO.	CLAIM, MINE, OR GROUP	OWNER NAME, ADDRESS	SEC.	T.	R.	B & M	REMARKS
	Mechanics	G. Allen Jr., Sutter Creek	4,5 32,33	6N 7N	11E 11E	MD MD	Active in 1895. North-trending vein with greenstone hanging wall and slate footwall, 240-ft. incline shaft with 100 and 200 levels. (Irelan 90:112; Crawford 96:74; Storms 00:76; Tucker 14:38; Logan 27:193; 34:120.)
	Middle Bar Q.M.	East Bay Municipal Utility District, Oakland	10	5N	11E	MD	See Littlefield.
	Middle Bar Group	East Bay Municipal Utility District, Oakland	3,10 15	5N	11E	MD	See Littlefield.
	Mikado						See Contini.
	Mineral Point	Domenico Boro, c/o Gene Boro, Jackson	3	5N	11E	MD	(Logan 27:193; 34:121.)
	Modoc	B.H. and Stella M. Mace, Pine Grove	21	7N	13E	MD	Active in 1895. (Crawford 96:74; Tucker 14:38; Logan 27: 193.)
	Monitor	Tony Levaggi et al. Plymouth, ¼, Chas. H. Shields et al., 580 Polk St., Apt. 2, Monterey, ½, D.J. Wilds et al. c/o Mayella Casinelli, Ione, ¼	13 14	7N	10E	MD	(Logan 27:193; 34:121.)
	Monte Richard	Ambrose and Guido Garbarini, c/o Babe Garbarini, Jackson	21 30	7N 6N	13E 11E	MD MD	
	Monte de Oro	Tony Levaggi, et al. Plymouth	26 35	8N	10E	MD	Massive quartz outcrops, 500-ft. open cut, 785-ft. tunnel, some activity in 1934. (Tucker 14:38; Logan 27:193; 34:...)

GOLD, LODE, (cont.)

MAP NO.	CLAIM, MINE, OR GROUP	OWNER NAME, ADDRESS	LOCATION				REMARKS
			SEC.	T.	R.	B & M	
	Monte de Oro (cont'd)						121.)
34	Moore	South Jackson Mining Co., c/o J.Schweitzer Jackson	34 33	6N	11E	MD	(Hanks 26:pt. II, 20; Irelan 88:84-85; Crawford 94:78; 96:74; Tucker 14:38; Logan 22:24; 23:299-300; 23:15-16; 24a:3; 24b:74; 24c:177; 27:173-175; Knopf 29:68-69; Logan 34:98-99; herein.)
	Mountain King, Mountain Queen	William B. Pitts, Pine Grove	4	6N	12E	MD	Series of north striking veins, auriferous galena present, 265-ft. shaft, 125-ft. shaft, drifts, and an adit. Active in early 1920's. A little activity in 1936. (Tucker 14: 38-39; Logan 24b:74-75; 27:193.)
	Muldoon Group	B. Monte Verda and E.C. Taylor, 369 Pine St. San Francisco	20	6N	11E	MD	(Logan 34:121.)
	Mungo Park	W.T. Kneen, Box 351 Mooreland, Oklahoma	20	7N	13E	MD	See Lone Willow Group.
	Nevada						See Bunker Hill.
35	Nevada Wabash	Nevada Wabash Mining Co., Box 504, Sutter Creek					(Logan 27:197; 34:123; herein.)
	New Albany Consolidated						See Littlefield.
36	New Hope	Harold and Emma M. Swingle, Plymouth	3	7N	10E	MD	(Irelan 90:121; Logan 27:194; herein.)
	New London	W.F. Detert Estate 1715 Mills Tower, San Francisco	14	7N	10E	MD	
	New London	B. Monte Verda and E.C.	14	7N	10E	MD	

GOLD, LODE, (cont.)

MAP NO.	CLAIM, MINE, OR GROUP	OWNER NAME, ADDRESS	SEC.	T.	R.	B & M	REMARKS
	New London (cont'd)	Taylor, 369 Pine St. San Francisco	11	7N	10E	MD	Now part of Plymouth Consolidated.
37	Newman	Pacific Gas and Electric Company, San Francisco	NE¼ 33	7N	13E	MD	(Logan 27:175; herein.)
	North California	W.F. Detert Estate 1715 Mills Tower, San Francisco	23	7N	10E	MD	(Logan 34:121.)
	North Eureka #2	W.F. Detert Estate 1715 Mills Tower, San Francisco	14	7N	10E	MD	(Logan 34:121.)
	North Gover (part of Fremont)		25	7N	10E	MD	(Irelan 90:116-117; Crawford 96:75; Logan 27:194; 34:121.)
	North Henry Clay	W.F. Detert Estate 1715 Mills Tower, San Francisco	14	7N	10E	MD	(Logan 34:121.)
	North Star	Nevada Wabash Mining Co., Box 504, Sutter Creek	6	6N	11E	MD	(Irelan 88:71-72; 90:99; Crawford 94:78; 96:75; Tucker 14:39; Logan 27:175-176; 34:99-101.)
	No. 1 and No. 2	Josephine Poundstone, Grimes, Colusa Co.	5,8	6N	11E	MD	(Logan 27:194; 34:121.)
38	Old Eureka						See Central Eureka.
	Old Oaker	P.A. and B.J. Southerland, 22 Linden Ave. Atherton	4	7N	10E	MD	(Logan 27:194; 34:121.)

GOLD, LODE, (cont.)

MAP NO.	CLAIM, MINE, OR GROUP	OWNER NAME, ADDRESS	SEC.	T.	R.	B & M	REMARKS
	Old Lone Willow	W.T. Kneen, Box 351 Mooreland, Oklahoma	20	7N	13E	MD	See Lone Willow Group.
	Oleta	C.D. Crane, et al., Fiddletown	34 17	8N 6N	11E 11E	MD MD	300-ft. shaft, some activity in middle 1920's. (Root 24:3; Logan 27:194.)
39	Oneida	South Eureka Mining Co c/o A.J. Mayman, 400 California St., San Francisco	17	6N	11E	MD	(Irelan 88:79; 90:109-110; Crawford 94:78; 96:75; Storms 00:60-63; Tucker 14:39; Logan 27:194; 34:121; Jenkins 48:64; herein.)
40	Original Amador	J.W. Bullock, c/o C.E. Crandall, North Bend Washington	36	7N	10E	MD	(Irelan 88:42; 90:114; Crawford 94:78; Tucker 14:39-40; Logan 24a:3; 27:178-179; Knopf 29:56-58; Logan 34:104-105 Jenkins 48:65; herein.)
	Oro Grande et al.	Marie R. North, 2008 Birch St., San Carlos	29 32 33	7N	13E	MD	(Logan 27:194.)
	Parker	Mrs. Elsie C. Downs Sutter Creek	14	7N	12E	MD	Several adits and a shaft, some activity in 1920's. (Logan 27:194.)
	Peerless (Seaton)	Hobart Estate Co. and Emma Rose, 202 Hobart Bldg., San Francisco	24	7N	10E	MD	Vein on slate-metavolcanic contact, inclined shaft. Active around 1900. (Storms 00:45; Logan 27:194.)
41	Peterson	W.F. Peterson, Pine Grove	5	6N	12E	MD	Herein.
	Pick and Drill	Bow Bridge Development Co., c/o E. Putman Box 1069, Greenwich Connecticut	32	7N	13E	MD	

GOLD, LODE, (cont.)

MAP NO.	CLAIM, MINE, OR GROUP	OWNER NAME, ADDRESS	SEC.	T.	R.	B & M	REMARKS
42	Pine Grove	Wendell M. Miller and E.A. DeRuchia, Reno Nevada	23	7N	13E	MD	Herein.
43	Pine Grove unit	W.F. Peterson, Pine Grove	32	7N	12E	MD	Herein.
	Pioneer	B. Monte Verda and E.C. Taylor, 369 Pine St. San Francisco	25	6N	11E	MD	See Argonaut.
	Pioneer	Glen and Everett Fancher, Jackson	14	7N	10E	MD	Two veins in Mariposa slates, 550-ft. inclined shaft, 5 levels, produced up to $15,000 per year in 1890's. (Irelan 88:42; 90:111-112; Crawford 94:78-79; 96:75; Storms 00:83-84; Tucker 14:41; Logan 27:195; 34:106.)
44	Pioneer Lucky Strike	J.H. Hauhuth, Pioneer Station	20	7N	13E	MD	(Crawford 94:78; 96:73; Tucker 14:37-38; Logan 27:172; herein.)
	Pitts	William B. Pitts, Pine Grove	4	6N	12E	MD	See Mountain King.
45	Plymouth Consolidated	B. Monte Verda and E.C. Taylor, 369 Pine St. San Francisco	11	7N	10E	MD	(Hanks 86: pt. II 15-16; Irelan 88:42-49; 90:117; 93:79; Crawford 94:79; 96:75; Storms 00:82-83; Tucker 14:37,41-43; Logan 21:411;22:24; 23:301; 23:16.73; 24a:3; 24c:177; 27:179-180; Knopf 29:6-7,49-51; Logan 34:106-111; Jenkins 48:65; herein.)
	Plymouth-Eureka	Richard Hosking, c/o B.H. Richardson, Box 387, Placerville	3	7N	10E	MD	In metavolcanics, 50-ft. shaft, active in 1914. (Tucker 14:43; Logan 27:195; 34:122.)
	Pocahontas	W.F. Detert Estate 1715 Mills Tower, San	24	7N	10E	MD	3 veins in slate, 620-ft. vertical shaft with 6 levels active in 1900. (Irelan 90:122-123; Storms 00:83; Tucker

GOLD, LODE, (cont.)

MAP NO.	CLAIM, MINE, OR GROUP	OWNER NAME, ADDRESS	LOCATION SEC.	T.	R.	B & M	REMARKS
	Pocahontas (cont'd) Francisco						14:44; Logan 27:195; 34:122.)
	Podesta	Julius Podesta, Jackson	21	6N	12E	MD	Quartz vein with pyrite and chalcopyrite in greenstone, worked intermittently by Bert Contini of Pine Grove.
	Price (Prize)	Tony Levaggi, Plymouth	26	8N	10E	MD	7-ft. vein with greenstone hanging wall and slate footwall, 4 shallow shafts. (Irelan 90:120-121; Logan 27:195; 34:122.)
	Providence	W.P. Detert Estate 1715 Mills Tower, San Francisco	14	7N	10E	MD	(Logan 35:122.)
	Quartz Mountain		19	7N	11E	MD	See Gold Mountain.
46	Rainbow	Claude Hanley, 1727 Hiawatha Ave., Stockton	32	7N	12E	MD	Herein.
	Red Cloud	K. Pierovich, Jackson	35	8N	10E	MD	6-ft. vein with greenstone foot wall, slate hanging wall 200-ft. ore shoot, 275-ft. shaft. (Irelan 90:119-120; Crawford 96:75-76; Storms 00:85; Tucker 14:45; Logan 27:195.)
47	Red Hill	Claude Hanley, 1727 Hiawatha Ave., Stockton	32	7N	12E	MD	Herein.
	Red Tape	Marie R. North, 2008 Birch St., San Carlos	32	7N	13E	MD	North trending vein in granodiorite, 110-ft. shaft, 500-ft. adit. Active in 1914. (Tucker 14:45; Logan 27:195.)
	Rhetta						See Amador Star.
	Rising Star	B. Monte Verda and E.C.	11	7N	10E	MD	Part of Plymouth Consolidated.

GOLD, LODE, (cont.)

MAP NO.	CLAIM, MINE, OR GROUP	OWNER NAME, ADDRESS	SEC.	T.	R.	B & M	REMARKS
	Rising Star (cont'd)	Taylor, 369 Pine St. San Francisco					
	Robinson						See Fort Ann.
	Sargent	East Bay Municipal Utility District, Oakland	10	5N	11E	MD	N. 10° W. vein in slate and diorite, 120-ft. inclined shaft 700 ft. of underground workings. (Irelan 90:107; Tucker 14:45; Logan 27:195; 34:122.)
	Schoolhouse						See Amador Gold.
	Seaton	Hobart Estate Co. and Emma Rose, 202 Hobart Bldg., San Francisco	24	7 N	10 E	MD	Massive quartz vein on slate-metavolcanic contact, 950-ft. inclined shaft. (Irelan 88:42; Crawford 96:76; Tucker 14:45; Logan 27:195; 34:122.)
	Shakespeare	John F. Davis Estate 2872 Jackson St., San Francisco	11	7 N	10 E	MD	Two veins in slate, 27-ft. shaft, 100-ft. adit. (Irelan 90:121-122; Logan 27:195; 34:122.)
	Somerset	Bow Bridge Development Co., c/o E. Putman, Box 1069, Greenwich, Connecticut	32	7N	13E	MD	
	South Cosmopolitan Group	W.F. Detert Estate 1715 Mills Tower, San Francisco	24	7 N	10 E	MD	Northwest trending vein on diorite-slate contact, 195-ft. adit, 310-ft. inclined shaft. Active in 1888. (Irelan 88:49; Logan 27:196; 34:123.)
48	South Eureka	South Eureka Mining Co. c/o A.J. Mayman, 400 California St., San Francisco	17	6N	11E	MD	(Irelan 90:113; 93:144; Crawford 94:79; 96:76-77;Storms 00:63-64; Tucker 14:45-47; Logan 27:180-182; Knopf 29:62-63; Logan 34:111-114; Jenkins 48:64; herein.)
	South Jackson	South Jackson Mining	28	6N	11E	MD	3 veins in slates and amphibolite schist, 577-ft. verti-

GOLD, LODE, (cont.)

MAP NO.	CLAIM, MINE, OR GROUP	OWNER NAME, ADDRESS	SEC.	T.	R.	B & M	REMARKS
	South Jackson (cont'd)	Co., c/o J. Schweitzer Jackson	.				cal shaft, 3 levels 2100 feet underground workings, much development work done in 1913-1914. (Tucker 14:47-48; Logan 27:196; 34:114.)
	South Keystone	Keystone Mining Co. c/o John D. Culbert Amador City	6	6N	11E	MD	Irelan 90:115; Storms 00:77; Logan 27:196; 34:114,123.)
49	South Spring Hill	Keystone Mining Co. c/o John D. Culbert Amador City	31	7N	11E	MD	(Hanks 86:18; Irelan 88:71,80-84; 90:73,98,99; 92:145; Crawford 94:79; 96:74,77; Storms 00:77; Tucker 14:48; Logan 27:182; 34:91,123; Jenkins 48:65; herein.)
	Spagnoli	Mrs. Dorothy Ferguson and D.S. Richmond, Waterman	9	6N	12E	MD	Some work done on tailings during 1938. (Storms 00:45; Logan 27:185,196.)
	St. Julian (St. Junian)	Anthony B. Caminetti Jackson	10	5N	11E	MD	(Crawford 96:77; Tucker 14:48; Logan 27:196; 34:135.)
	Sunset		9	5 N	11 E	MD	See Argonaut. (Root 24:75; Logan 27:196.)
	Tellurium	Dr. F. O'Neil, Oroville and C.Brockman West Point	33	7 N	12 E	MD	Some work done during 1934 and 1945. (Crawford 94:79; Crawford 96:77; Tucker 14:49; Logan 27:197; 34:123.)
	Templeton	Mrs. Mary Templeton c/o Mrs. J. Gabs, 1112 Pacific Ave., Alameda	2	7N	10E	MD	
	Tom and Dick	Bow Bridge Development Co., c/o E. Putman, Box 1069, Greenwich, Connecticut	32	7N	13E	MD	

GOLD, LODE, (cont.)

MAP NO.	CLAIM, MINE, OR GROUP	OWNER NAME, ADDRESS	LOCATION				REMARKS
			SEC.	T.	R.	B & M	
50	Treasure	Treasure Mining Co. 220 Montgomery St. San Francisco	25	7 N	10 E	MD	(Crawford 96:77-78; Tucker 14:49-50; Logan 21:411-412; 23:300; 23:16; 27:183-184,197; 34:123,199; Jenkins 48:65; herein.)
	Trench	West Point Consolidated Mines Co., c/o E.H. Outerbridge, 250 Park Ave., New York, New York	32	13N	7E	MD	
	Trencher	West Point Consolidated Mines Co., c/o E.H. Outerbridge, 250 Park Ave., New York, New York	32	13N	7E	MD	
	Tri Mountain	The Knight Company Sutter Creek	22	7N	12E	MD	
	Triumph	Helen Crotty, Sutter Creek	6	6N	11E	MD	See Golden Eagle.
51	Valpariso	Valpariso Mining Co. Jackson	10	5N	11E	MD	Irelan 88:42; Crawford 94:79; 96:78; Tucker 14:52; Logan 27:197; 34:114-115,123; herein.)
	Victoria	Richard D. Dixon et al 425-4th St., Santa Rosa	11,24	7N	10E	MD	(Logan 34:123.)
	Volunteer	B. Monte Verda and E.C. Taylor, 369 Pine St. San Francisco	20	6N	11E	MD	(Irelan 90:105; Logan 27:197; 34:123.)
	Wabash	Nevada Wabash Mining Co., Box 504, Sutter	6	6N	11E	MD	(Logan 27:197; 34:123.)

GOLD, LODE, (cont.)

MAP NO.	CLAIM, MINE, OR GROUP	OWNER NAME, ADDRESS	LOCATION					REMARKS
			SEC.	T.	R.	B & M		
	Wabash (cont'd)	Creek						
	West Eureka		7	6N	11E	MD		See Central Eureka. (Tucker 14:52; Logan 27:197; 34:123.)
	West Somerset	West Point Consolidated Mines Co., c/o E.H. Outerbridge, 250 Park Ave., New York New York	29	7N	13E	MD		
	White Mountain	Harbison and Walker Mining Co., 1800 Farmers Bank Bldg., Pittsburg, Pennsylvania	19	7N	11E	MD		(Logan 27:197; 34:123.)
52	Wildman							See Lincoln Consolidated.
	Wonder	Keystone Mining Company, c/o John D. Culbert, Amador City	36	7N	10E	MD		Herein.
53	Zeila (Zeile)	Mark and Frances Eudey, Martell	28	6N	11E	MD		(Hanks 86:22-23; Irelan 88:22,68,85,88; 90:68,104; 92:139; Crawford 94:80; 96:79; Storms 00:45,50; Tucker 14:52; Logan 27:184-185,197; 34:115-116, 123; Jenkins 48:62; herein.)

GOLD, PLACER

MAP NO.	CLAIM, MINE, OR GROUP	OWNER NAME, ADDRESS	LOCATION SEC.	T.	R.	B & M	REMARKS
	Amador Dredging Co.	Amador Dredging Co., Ione					Operated dragline dredge in Ione district part of 1941. (Averill 46:231.)
	Arroyo Seco Gold Dredging Company	Arroyo Seco Gold Dredging Co., 351 California St., San Francisco					Operated electric connected-bucket dredge from Jan. 1 to May 15, 1941. Also operated throughout 1940 at property 3 miles west of Ione and in 1935. (Averill 46:231.)
	Carlson and Sandburg						Operated dragline dredge east of the Ione city limits in 1939.
	Columbus Gold Gravel (Drift)	Charles W. Cassinelli et. al., c/o A. Marrie 643 Chetwood St., Oakland	9	7N	12E	MD	Worked deposit 1 mile north of Mokelumne River in Camanche district in the fall of 1935.
	Comanche Gold Dredging Company						Operated dragline dredge on South Fork of Consumnes River 1937-39.
	Cuffe, Frank						Operated dragline dredge on South Fork of Consumnes River 1937-39.
54	Delta Placer Gold	Glen O'Brien, Ione	10	5N	9E	MD (proj.)	Herein.
	Diebold Ranch						L.E. Pearson of San Francisco operated a dryland dredge near Camanche in 1942.
55	Elephant hydraulic		15	7N	12E	MD	1½ to 3 feet of gravel on decomposed slate bedrock overlain by 40 to 45 feet of volcanic ash. Worked in 1922-23, 1927, and 1932. (Logan 27:198; Averill 46:231.)
	Flower Gold Recovery Co.						Operated dragline and dryland plant near Ione in 1938-39.

GOLD, PLACER, (cont.)

MAP NO.	CLAIM, MINE, OR GROUP	OWNER NAME, ADDRESS	SEC.	T.	R.	B & M	REMARKS
56	Garibaldi dredge	Frank and Pete Garibaldi, Volcano	23	7N	12E	MD	(Averill 46:231; herein.)
	Gold Hill Dredging Co.						Operated 5 cubic ft. dredge with 150,000 yards per month capacity in 1935 at Frank Collins property in Arroyo Seco
	Henry and Weaver	Allen Ranch property					Operated dragline dredge in Sutter Creek gulch part of 1941. (Averill 46:231.)
	Horseshoe Dredging Co.	Horseshoe Dredging Co. Ione					Operated dragline dredge in Ione district May to July 1940. (Averill 46:231.)
	Horton	H.G. Kreth, Ione					Hydraulicked Horton mine in Jackson Valley, 5 miles south of Ione during parts of 1941. (Averill 46:231.)
	Humphreys Gold Corporation	Humphreys Gold Corporation, Denver, Colorado					Operated dry-land dredge near Carbondale in 1938.
	Irish Hill	McQueen and Downing, 1040 38th Street, Sacramento					Operated dragline dredge from March 28 to June 25, 1941. (Averill 46:232.)
	Independence Gold Mines	Independence Gold Mines					Operated stationary washing plant in Camanche district between July 30 and October 12, 1941. (Averill 46:231.)
	Kate Gray	Mary and Albina Delucchi, Volcano	14	7N	12E	MD	
	Kent Dredging Company	E.A. Kent, 351 California St., San Francisco					Operated two dragline dredges on Sutter Creek between Volcano and Sutter Creek during 1940. (Averill 46:232.)
	Lagomarsino drift tunnel	D. Hendrickson, Oakland (lessee)					Some work done in late 1933 and early 1934.

GOLD, PLACER, (cont.)

MAP NO.	CLAIM, MINE, OR GROUP	OWNER NAME, ADDRESS	SEC.	T.	R.	B & M	REMARKS
	Lancha Plana Gold Dredging Company	Lancha Plana Gold Dredging Co., La Lomita Rancho, Lockeford					Operated electric connected-bucket dredge on Jackson Creek near Buena Vista throughout 1940 and Jan. to May, 1941. Also worked on Stony Creek half way between Ione and Jackson during 1937.
57	Lilly dredge	E.L. Lilly, 1640 East Poplar St., Stockton 5	24	8N	9E	MD	(Logan 46:233; herein.)
	Long Bar Gold Dredging Company	Lorentz property	20,28, 29	8N	10E	MD	Operated dragline dredges on Consumnes River in 1942. (Averill 46:232.)
	Lorentz and Swingle	Lorentz property					Operated on Consumnes River; 7 miles northwest of Plymouth in 1941. (Averill 46:232.)
58	Madrill Dredge	C.R. and Annabelle C. Brown, R.F.D., Ione	6	4N	10E	MD	Herein.
	McDonald dredge						Operated dragline dredge on Dry Creek in August 1949.
	Mountain Gold Dredging Company	Mountain Gold Dredging Company, Plymouth					Operated dragline dredge on Matulich property near Drytown intermittently during 1941. Also operated on Detert Estate in Plymouth district on Indian Creek in 1941. (Averill 46:232.)
	Orr, Elwood						Elwood Orr was in charge of a placer operation 3 miles north of Plymouth in the middle part of 1936. Treated between 200 to 250 cubic yards per day.
	Pacific Placers Engineering Company	Pacific Placers Engineering Co., Sacramento					Operated dragline dredge on McColloh property in Ione district during parts of 1940-41. (Averill 46:232.)
	Pantle, J.C.	J.C. Pantle, Lincoln					Operated dry-land dredge on Willow Creek 5 mi. west of Drytown from 1940 to Oct., 1942. (Averill 46:232.)

GOLD, PLACER, (cont.)

MAP NO.	CLAIM, MINE, OR GROUP	OWNER NAME, ADDRESS	SEC.	T.	R.	B & M	REMARKS
59	Pigeon drift	Fred N. Pigeon, Fiddletown	28	8N	11E	MD	(Tucker 14:41; herein.)
	Placeritas Mining Co.	Placeritas Mining Co. 245 N. Gramercy Place Los Angeles					Operated dragline dredge on six different properties within 4 mile radius of Plymouth. (Averill 46:232.)
	Plymouth Gold Dredging Company	Plymouth Gold Dredging Company, Plymouth					Operated dryland plant in Plymouth area during part of 1940.
	R. & M. Mining Company	R. & M. Mining Company Plymouth					Operated dragline dredge, beside State Highway 49, 2 mi. north of Plymouth in 1938.
	Raymond Mining Co., 114 N. Westmoreland Ave., Los Angeles 4.	Frank P. & Annie Dal Porto, Plymouth	SW¼ SE¼ 6 NW¼ NE¼ 7	7N	11E	MD	This company leased gravel beds along N. Fork of Dry Creek in Feb. 1952.
	Rendell	Lee and Lillie A. Payton, Pine Grové	35	7N	12E	MD	Operated dragline dredge in Ione district from Feb. 4 to May 26, 1941. Also operated near Drytown on Consumnes River between Plymouth and Nashville in 1940. (Averill 46:232,233.)
	Rim Cam Gold Dredging Co.	Yager Ranch					Operated dragline dredge near Aukum from Jan. to June 12, 1941. This was the most productive dragline dredge in the county in 1940. (Averill 46:232.)
	River Pines Mining Co.	River Pines Mining Co. Plymouth					Operated dragline dredge on Arroyo Seco Ranch during part of 1940. (Averill 46:232.)
	San Andreas Gold Dredging Company	San Andreas Gold Dredging Co., 960 Russ Bldg. San Francisco					

GOLD, PLACER, (cont.)

MAP NO.	CLAIM, MINE, OR GROUP	OWNER NAME, ADDRESS	LOCATION				REMARKS
			SEC.	T.	R.	B & M	
60	Secor dredge	Evans, Elmer, c/o County Treasurer's Office, Jackson	18	7N	12E.	MD	(Herein.)
	Stock, Lee						Operated dragline dredge on Dry Creek on Rancho Arroyo Seco between Waits Station and Ione in 1949. The outfit handled 70 cubic yards per hour.
	Treble Clef	E.L. Lilly, 706 California Bldg., Stockton					Operated dragline dredge during parts of 1941. This outfit was operated throughout 1940 on 2 different properties. (Averill 46:233.)
	Union Flat	Chester Bonneau Charles Guilian and James Giannini					Property worked during 1927. Some hydraulicking was done under lease during 1938. Some clean-up work was done in 1936. Lagomarsino hydraulicked property from March through May 1949. (Logan 27:199.)

IRON

MAP NO.	CLAIM, MINE, OR GROUP	OWNER NAME, ADDRESS	LOCATION				REMARKS
			SEC.	T.	R.	B & M	
61	Arroyo Seco		27 28	6N	9E	MD (proj.)	(Aubury 06:364; Tucker 14:52; Logan 27:199; herein.)
62	Clinton (Irishtown)		8,9	6N	12E	MD	(Aubury 06:364; Tucker 14:52; Logan 27:199; herein.)
63	Thomas	George M. Thomas, Jackson	16 17	5N	10E		Herein.

MANGANESE

MAP NO.	CLAIM, MINE, OR GROUP	OWNER NAME, ADDRESS	LOCATION SEC.	T.	R.	B & M	REMARKS
	Alexander	Darrow A. Russell, et al., Amador City	29	7N	11E	MD	Bed of manganiferous and ferruginous chert in slate. (Jenkins 43:75; Trask 50:27)
	Big Gulch	State of California	NW¼ 4	7N	12E	MD	Old hydraulic gold property. Low-grade manganese oxide in lateritic soils beneath Tertiary lava cap. (Jenkins 43:75; Trask 50:27.)
	Custer (Dooley)	R.B. Custer, et al., c.o R.J. Custer, Route 2, Box 1077, Stockton	28	6N	10E	MD	Manganiferous metachert enclosed in metavolcanics and metatuffs. (Jenkins 43:75; Trask 50:27-28.)
	Deaver	E.E. Hutchinson, et al. 207 Union Street, Placerville	31	8N	12E	MD	A few blocks of manganiferous metachert exposed in road-cut. (Trask 50:28.)
64	DuFrene	Fred DuFrene, et al. Ione	SW¼ 27	6N	10E	MD	A bed of manganiferous metachert five to eight feet thick is exposed for a length of 270 feet. Worked by small shaft and series of open cuts. (Jenkins 43:75; Trask 50:28-29.)
	Eagle's Head	State of California	N½ 4	7N	12E	MD	Concentrations low-grade manganese oxide in lateritic soils beneath Tertiary lava cap. (Jenkins 43:75; Trask 50:29.)
	Germolis (Rodonick)	Paul Germolis, et al. Riddletown	NW¼ 10	7N	11E	MD	A development cut exposes bed of massive, manganiferous metachert containing pockets 3-4 inches wide of high-grade ore. The average analysis for the bed is probably around 20% Mn. (Jenkins 43:75; Trask 50:30.)
	Jones	D.C. Stowe, 620 East Charter Way, Stockton	32	7N	12E	MD	Manganiferous metachert. The ore is chiefly botryoidal and nodular psilomelane with some pyrolusite worked by 30-foot shaft and open cuts. Last work was World War I but no ore shipped. (Trask 50:30-31.)

MANGANESE, (cont.)

MAP NO.	CLAIM, MINE, OR GROUP	OWNER NAME, ADDRESS	LOCATION				REMARKS
			SEC.	T.	R.	B & M	
	Lubanko	Louis Lubanko, Fiddletown	SE¼ 10	7N	11E	MD	(Jenkins 43:75; Trask 50:31; herein.)
65	Perini (Peyton et al. lease)	Benjamino Perini, Pine Grove	NW¼ 35	7N	12E	MD	(Bradley et al., 18:30; Logan 27:201; Trask 50:31-32; herein.)
66	Peyton (Crocker-Preston property)	Benjamino Perini, Pine Grove	SW¼ 35	7N	12E	MD	(Bradley et al., 18:29; Logan 27:200; Trask 50:32-33; herein.)
	Ruhser and Hubberty		29?	7N	13E	MD	(Bradley et al., 18:30; Logan 27:201; Jenkins 43:75; Trask 50:33.)
	Stirnaman	J.L. and L.C. Wells 716 Wagner Avenue Stockton	SE¼ 24	7N	12E	MD	(Logan 27:201; Jenkins 43:75; Trask 50:34-35; herein.)

COAL

| MAP NO. | CLAIM, MINE, OR GROUP | OWNER NAME, ADDRESS | LOCATION | | | | REMARKS |
			SEC.	T.	R.	B & M	
71	American Lignite Products Co.	Maud May Fithian (1945) (Operated by American Lignite Products Co., Ione)	SE¼ 26	6N proj	9E	MD	(Jenkins 49:35; 50:62; 51:311; herein)
72	Buena Vista	Leased by American Lignite Products Co., Ione	NE¼ 19	5N	10E	MD	Last period of continuous operation was from 1925 to 1938. Coal was mined by underground methods off two shafts. (Tucker 14:11; Logan 21:413, 27:146)
73	Carbondale	Charles S. Howard Estate, c/o Crocker First National Bank of San Francisco, Post and Montgomery Sts., San Francisco	NE¼ 5	6N	9E	MD	At Carbondale. Lot 324, Rancho Arroyo Seco. (Irelan 92: 148) Leased to Gladding, McBean and Co., 2901 Los Feliz Blvd., Los Angeles 26, from 1948 to date.
74	Coal Mine No. 1	Charles S. Howard Estate, c/o Crocker First National Bank of San Francisco, Post and Montgomery Sts., San Francisco	NE¼ 35(?)	6N proj	9E	MD	Lot 259, Rancho Arroyo Seco. Leased to Gladding, McBean and Co., 2901 Los Feliz Blvd., Los Angeles 26, from 1948 to date.
75	Coal Mine No. 2	Charles S. Howard Estate, c/o Crocker First National Bank of San Francisco, Post and Montgomery Sts., San Francisco	S½ 26	6N	9E	MD	Lot 437, Rancho Arroyo Seco. Leased to Gladding, McBean and Co., 2901 Los Feliz Blvd., Los Angeles 26, from 1948 to date.
76	Coal Mine No. 3	Charles S. Howard Estate, c/o Crocker First National Bank of San Francisco, Post and	NE¼ 16	6N proj	9E	MD	Lot 237, Rancho Arroyo Seco. 0.5 of a mile northwest of Clarksons. (Irelan 88:110-112; 92:147; Crawford 94:41, 96:51; Logan 27:148)

COAL, (cont.)

MAP NO.	CLAIM, MINE, OR GROUP	OWNER NAME, ADDRESS	SEC.	T.	R.	B & M	REMARKS
76	Coal Mine No. 3 (cont'd)	Montgomery Sts., San Francisco. Leased to Gladding, McBean and Co., 2901 Los Feliz Blvd., Los Angeles 26, from 1948 to date					
77	Coal Mine No. 4	Charles S. Howard Estate, c/o Crocker First National Bank of San Francisco, Post and Montgomery Sts., San Francisco. Leased to Gladding, McBean and Co., 2901 Los Feliz Blvd., Los Angeles 26, from 1948 to date	SW¼ 4	6N proj	9E	MD	Lot 232, Rancho Arroyo Seco. 1 mi. E. of Carbondale ¼ mile N. of RR. (Tucker 14:12; Logan 27:147)
78	Harvey		NE¼ 32	7N	9E	MD	1 mi. N. of Carbondale at May. (Tucker 14:11; Logan 21: 413; 27:147)

CRUSHED ROCK, SAND & GRAVEL

MAP NO.	CLAIM, MINE, OR GROUP	OWNER NAME, ADDRESS	SEC.	T.	R.	B & M	REMARKS
	Ayers, Charles, Sutter Creek	Central Eureka Mining Company, Russ Building San Francisco	8	6N	11E	MD	Dry bar run stream gravels from Sutter Creek have been produced intermittently since early 1930's. Last production was during 1948. The amount and time of production depend upon the market. A ¼ yard dragline is on the property.
68	Alpine Gravel Plant	Jack Bacigalupi, Pine Grove	NE¼ SW¼ 22	6N	11E	MD	Herein.
69	Relvas gravel	E.J. Relvas, Ione	14,15	6N	9E	MD (proj.)	Herein.
70	Sacramento County sand pit	Louis G., Cora J., and Gene L. Klotz, Freeport Blvd., Rt. 8, Box 1380, Sacramento 17	SW¼ 5	7N	9E	MD	Herein.

LIMESTONE & MARBLE

MAP NO.	CLAIM, MINE, OR GROUP	OWNER NAME, ADDRESS	LOCATION				REMARKS
			SEC.	T.	R.	B & M	
79	Allen Estate	John B. and George Allen, Sutter Creek	11	6N	10E	MD	Of a number of limestone outcrops, only few show promise. One of outcrops worked in late 1945 by A. Teichert & Son, Inc. of Sacramento (Analysis herein).
	Amador Marble						See Dal Porto marble. (Aubury 06:96; Tucker 14:53; Logan 27:200; 47:209)
	Amador Lime Rock	Pacific Portland Cement Co., 417 Montgomery St., San Francisco	16	6N	10E	MD	Worked in small way from 1859 to 1910; shallow pits on property. Limestone is light gray, compact, finely crystalline. (Aubury 06:64; Logan 47:208)
	Garibaldi Ranch	Theresa Garibaldi, Amador City	N½ 33	7N	10E	MD	Main limestone outcrop is 120 feet wide and 930 feet long. High magnesian content. No work has been done. (Logan 47:210; analysis herein)
80	Dal Porto Marble	Frank P. Dal Porto, Plymouth	S½ SW¼ and SW¼SE¼ 6	7N	11E	MD	Worked in 1880 or earlier producing marble. Property has quarry face 90 feet, waste dump 200 feet long and 40 feet wide. Best marble 50 feet wide in center with white to dark gray color. (Aubury 06:97; Tucker 14:53; Logan 27:200, Logan 47:209; analysis herein)
81	Dondero Marble	Aurelio G. Dondero, 852 59th St., Oakland	N½ 29	7N	12E	MD	(Aubury 06:96-97; Tucker 14:53; Logan 27:200, 47:209; analysis herein)
	Ellis Bros. Ranch	J.M. and F.J. Ellis, Box 98, Jackson	30	6N	11E	MD	No work done; several limestone outcrops.
	Fiddletown	Pacific Portland Cement Co., 417 Montgomery St., San Francisco	5,6 7	7N	12E	MD	Large and irregular limestone outcrops. Deposit partly covered by andesite and gravel, partly by soil. Undeveloped. (Logan 47:210; analysis herein)
	Grelich Ranch (Grelich Ranch)	Geo. E. & Eleanor Grelich, Plymouth	5,8	7N	10E	MD	Main limestone outcrop 450 feet long, 50 to 85 feet wide. Limestone is finely crystalline and gray in color. Undeveloped. (Logan 47:211; analysis herein)

LIMESTONE & MARBLE, (cont.)

MAP NO.	CLAIM, MINE, OR GROUP	OWNER NAME, ADDRESS	LOCATION				REMARKS
			SEC.	T.	R.	B & M	
	Oleta marble						See Dal Porto marble.
82	Volcano Limestone	Great Western Lands Development Co. c/o J.J. Lerman, 806 Balboa Bldg. San Francisco	23	7	12	MD	Six diamond drill holes drilled 1928. Probably largest high-calcium, low-magnesium limestone in Amador Co. Undeveloped (Logan 27:200, 47:211; analyses herein)
	Wait Marble	Greilich Bros., Drytown	21	7	10	MD	Marble is hard, durable, and takes a high polish, having cherry red color due to iron oxides. It crops out over area 160 x 400 feet. One hole drilled 1924. (Logan 27:300 47:212)

PUMICE

MAP NO.	CLAIM, MINE, OR GROUP	OWNER NAME, ADDRESS	LOCATION				REMARKS
			SEC.	T.	R.	B & M	
67	Bacon Pit	Charles S. Howard Estate, c/o Crocker First National Bank San Francisco	SE¼ 19	6N	9E (proj.)	MD	(Jenkins 50:193; herein.)

REFRACTORY MATERIALS
CLAY

MAP NO.	CLAIM, MINE, OR GROUP	OWNER NAME, ADDRESS	SEC.	T.	R.	B & M	REMARKS
83	Airplane	Charles S. Howard Estate, c/o Crocker First National Bank of San Francisco, Post and M Montgomery Sts., San Francisco. Leased to Gladding McBean and Co., 2901 Los Feliz Blvd., Los Angeles 26, from 1948 to date.	NW¼ 15	6N	9E (proj)	MD	(Bates 45:6,24; Johnson 48:1-6; 49:491-498; herein.)
84	Amick	State of California, Preston School of Industry, Waterman	23 24 (?)	6N	9E (proj)	MD	Half a mile west of Ione. Formerly operated by W.D. Amick, Ione, and Philadelphia Quartz Company, Berkeley. Ione sand from the lower member of the Ione formation. Sandy clay was washed to produce china clay and glass sand. Open pit. Idle. (Tucker 14:9-10; Boalich 20:39-42; Logan 21:413, 23:95; Dietrich 28:63; Sampson and Tucker 31:433-34.)
	Bacon Red						See Ione Red.
85	Baker Hill	Charles S. Howard Estate, c/o Crocker First National Bank of San Francisco, Post and Montgomery Sts., San Francisco	NE¼ 9	6N	9E (proj)	MD	Lot 234, Rancho Arroyo Seco. 4½ miles N. 45° W. of Ione. Fire clay (P.C.E.: cone 33-34) from the lower member of the Ione formation. Ceased operation about 1944. Open pit. Idle. (Tucker 14:11; Logan 27:135; Dietrich 28:pl. 6)
	Barber						See Shepard.
	Carlile clay and sand						See Solari.
86	Chalk Hill	Charles S. Howard Estate, c/o Crocker First National Bank of San Francisco, Post and	SE¼ 10	6N	9E (proj.)	MD	Lots 222 and 223, Rancho Arroyo Seco. 3 miles N. 40° W. of Ione. Fire clay (P.C.E.: cone 33) from the lower member of the Ione formation. Similar to Cheney Hill clay. Open pit. Idle. (Bates 45:6, 24)

CLAY,(cont.)

MAP NO.	CLAIM, MINE, OR GROUP	OWNER NAME, ADDRESS	SEC.	T.	R.	B & M	REMARKS
86	Chalk Hill (cont'd)	Montgomery Sts., San Francisco. Leased to Gladding, McBean and Co., 2901 Los Feliz Blvd., Los Angeles 26, from 1948 to date.					
87	Cheney Hill	Charles S. Howard Estate, c/o Crocker First National Bank of San Francisco, Post and Montgomery Sts., San Francisco. Leased to Gladding, McBean and Co., 2901 Los Feliz Blvd., Los Angeles 26, from 1948 to date	SW¼ 35	6N	9E (proj.)	MD	(Bates 45:5-6, 24-26; herein)
88	China Gulch		NW¼ 32	5N	10E	MD	On the west side of China Gulch 1.8 miles N. 16° W. from Lancha Plana. Fire clay (P.C.E.: cone 33) from the lower member of the Ione formation. Cheney-Hill-type clay according to Bates but possibly Edwin-type clay. Open pit and adits. Idle. (Bates 45:5-6, 24)
89	Chitwood	C.W. Chitwood (1945)	SE¼ 13	5N	9E (proj.)	MD	1.2 miles S. 55° W. from Buena Vista. Chitwood clay from the upper member of the Ione formation. A gray sandy clay with a P.C.E. of cone 30-31. Open pit. Idle. (Bates 45: 6, 24; Pask 52:21,23)
90	Chocolate (Harvey)	M.J. Bacon, Ione (1928)	NE¼ 32 (?)	7N	9E	MD	Apparently the pit 0.6 mile north of Carbondale. Clay from the lower member of the Ione formation. Open pit. Idle. (Logan 27:135; Dietrich 28:57, 280)
	Clark						(See Dosch)

CLAY, (cont.)

MAP NO.	CLAIM, MINE, OR GROUP	OWNER NAME, ADDRESS	SEC.	T.	R.	B & M	REMARKS
91	Clark Sand	Pacific Clay Products Co. Box 145, Sta. A, Los Angeles 31	SW¼ 28	7N	9E	MD	(Aubury 06:208; Tucker 14:6; Logan 27:136; Dietrich 28: 58, 261; Sampson and Tucker 31:435; herein.)
92	Deutschke	Charles J. Deutschke Ione, (1945)	N½ 21	6N	9E (proj.)	MD	Herein.
93	Dosch (Clark)	Pacific Clay Products Co. Box 145, Sta. A., Los Angeles 31	N½ 15	6N	9E (proj.)	MD	(Irelan 88:108; Tregloan 03:13-14; Aubury 06:207-208; Tucker 14:5-6; Boalich 20:38; Logan 23:96, 27:136; Dietrich 28:58, 302,312; Allen 29, 359; Bates 45:6, 24; herein.)
94	Edwin (Jones Butte) Pit No. 1 Pit No. 2 Pit No. 3 Pit No. 4	Charles S. Howard Estate, Crocker First National Bank of San Francisco, Post and Montgomery Sts., San Francisco. Leased to Gladding, McBean and Co., 2901 Los Feliz Blvd., Los Angeles 26, from 1944 to date.	S½ 16 N½ 21	6N	9E (proj.)	MD	(Logan 27:141-42; Dietrich 28:53-54; Schuette 29:196-198; Bates 45:1-38; herein.)
	Mine No. 1 Mine No. 2 Mine No. 3 Mine No. 4						
95	Fancher	J.M. Fancher, Ione	SE¼ 12	5N	9E (proj.)	MD	1 mile N. 77° W. of Buena Vista. Fire clay from the Cheney Hill bed in the lower member of the Ione formation. Open pit and adits. Idle. (Logan 23:96; Logan 27:141; Dietrich 28:59,280; Bates 45:6, 24; Pask 52:17,19,23)
96	Farci	Antoni Farci, Ione. Leased to Western Refractories Co., Russ Bldg., San Francisco since 1950	SW¼ 5	5N	10E	MD	Herein.

CLAY, (cont.)

MAP NO.	CLAIM, MINE, OR GROUP	OWNER NAME, ADDRESS	SEC.	T.	R.	B & M	REMARKS
	Free						See King and Enos.
	Gage						See Irish Hill.
	Hammer	Fred Hammer, Carbon-dale (1896)					At Carbondale. Mined clay for use in the Carbondale Pottery at May and to sell. Operated from 1877 through 1896. (Watts 93:149; Crawford 96:613.)
	Harvey						See Yosemite and Chocolate.
97	Ione Fire Brick	Charles S. Howard Estate, c/o Crocker First National Bank of San Francisco, Post and Montgomery Sts., San Francisco. Leased to Gladding, McBean and Co., 2901 Los Feliz Blvd., Los Angeles 26, from 1948 to date.	NW¼ 36	6N (proj.)	9E	MD	Lot 273, Rancho Arroyo Seco. 1 mile south of Ione and just west of the Shepard pit. Clayey sand from the Ione Sand bed in the lower member of the Ione formation. Open pit. Active in 1928. Now idle. (Logan 27:138-141; Dietrich 28:56,280)
	Ione Fire Brick Company Plant						See Western Refractories Company Plant.
98	Ione Red (Bacon Red) (Lane Mottled)	Mrs. M. J. Bacon, Ione Leased to Gladding, McBean and Co., 2901 Los Feliz Blvd., Los Angeles 26, from 1948 to date.	NW¼ 32	6N	10E	MD	(Logan 27:135; Dietrich 28:57,335; Bates 45:6,24; herein)
99	Ione Sand	Charles S. Howard Estate, c/o Crocker First National Bank of San	N½ 36	6N (Prbj.)	9E	MD	Herein.

CLAY, (cont.)

MAP NO.	CLAIM, MINE, OR GROUP	OWNER NAME, ADDRESS	SEC.	T.	R.	B & M	REMARKS
99	Ione Sand (cont'd)	Francisco. Leased to Gladding, McBean and Co., 2901 Los Feliz Blvd., Los Angeles 26, from 1948 to date.	N½ 36	6N	9E	MD (proj.)	Herein.
100	Irish Hill (Gage)	Charles S. Howard Estate, c/o Crocker First National Bank of San Francisco, Post and Montgomery Sts., San Francisco. Leased to Gladding, McBean and Co., 2901 Los Feliz Blvd., Los Angeles 26, from 1948 to date.	NE¼ 10	6N	9E	MD (proj.)	Lot 224, Rancho Arroyo Seco. 3.9 miles N. 31° W. of Ione. Residual clay in the Mariposa slate. Open pit. Idle. (Aubury 06:208-210; Tucker 14:11; Logan 27:135; Dietrich 28: 52-53,273).
	Jones Butte						See Edwin.
101	Kaolin-Fye	Calaveras Cement Company, 315 Montgomery St., San Francisco 6 and Pearl T. Fye. Leased to Calaveras Cement Company	SW¼ 8	5N	10E	MD	(Pask 52:19,23; herein)
102	King and Enos (Free)	Charles S. Howard Estate, c/o Crocker First National Bank of San Francisco, Post and Montgomery Sts., San Francisco. Leased to Gladding, McBean and Co., 2901 Los Feliz	N½ 5	6N	9E (proj.)	MD	Lot 324, Rancho Arroyo Seco. 200 yards west of Carbondale. Soft, plastic, light blue, refractory clay, with no sand, in the lower member of the Ione formation. Clay bed from 6 to 14 feet thick. Average annual production in 1906 was about 9000 tons. Open pit. Idle. (Watts 93:149; Tregloan 03:13-14; Aubury 06:208; Tucker 14:11)

CLAY, (cont.)

MAP NO.	CLAIM, MINE, OR GROUP	OWNER NAME, ADDRESS	SEC.	T.	R.	B & M	REMARKS
102	King and Enos (Free) (cont'd)	Blvd., Los Angeles 26, from 1948 to date.					
	Lane mottled						See Ione Red.
103	Laterite	Charles S. Howard Estate, c/o Crocker First National Bank of San Francisco, Post and Montgomery Sts., San Francisco. Leased to Gladding, McBean and Co., 2901 Los Feliz Blvd., Los Angeles 26, from 1948 to date.	W½ 16	6N (proj.)	9E	MD	(Dietrich 28:54; Allen 29:383-394; Bates 45:13-15; herein)
104	Newman	Nettie V. Gradwohl (1945). Leased to Western Refractories Co., Russ Bldg., San Francisco	NW¼ 31, SW¼ 30	6N	10E	MD	1.0 mile S 18° E. of Ione on both sides of the Amador Central railroad. Clayey sand from the Ione Sand bed in the lower member of the Ione formation. Average annual production during the 1920's about 6000 tons. Open pit and room and pillar stopes. Idle since 1934. (Tucker 14:7-9; Boalich 20:38; Logan 23:96; Logan 27:140-141; Dietrich 28:61-63, 261,290,335; Allen 29:440; Sampson and Tucker 31:435)
105	Newman Red	Nettie V. Gradwohl (1945). Leased to Western Refractories Co., Russ Bldg., San Francisco	NW¼ 31	6N	10E	MD	(Logan 27:141; Dietrich 28:63,329; Bates 45:24; herein)
	Preston Brick plant	State of California, Preston School of Industry, Waterman	24	6N (proj.)	9E	MD	Several pits on state property. Intermittent production for use in common brick. Open pit. Currently idle.

CLAY, (cont.)

MAP NO.	CLAIM, MINE, OR GROUP	OWNER NAME, ADDRESS	SEC.	LOCATION			REMARKS
				T.	R.	B B M	
106	Shepard (Barber)	Charles S. Howard Estate, c/o Crocker First National Bank of San Francisco, Post and Montgomery Sts., San Francisco. Leased to Gladding, McBean and Co., 2901 Los Feliz Blvd., Los Angeles 26, from 1948 to date.	NE¼ 36	6N (proj.)	9E	MD	Lot 261, Rancho Arroyo Seco. 1.0 mile S. 6° E. of Ione. Adjoins Newman pit on the west. Buff clayey sand about 16 feet thick from the Ione sand bed in the lower member of the Ione formation. Overburden of brown clayey sand 6 to 10 feet thick. Open pit and room and pillar stopes. Idle. (Logan 27:135-36; Dietrich 28:54-56, 261; Sampson and Tucker 31:434)
107	Smokey No. 1	Charles S. Howard Estate, c/o Crocker First National Bank of San Francisco, Post and Montgomery Sts., San Francisco. Leased to Gladding, McBean and Co., 2901 Los Feliz Blvd., Los Angeles 26, from 1948 to date.	SW¼ 27	6N (proj.)	9E	MD	Lot 285, Rancho Arroyo Seco. 2.6 miles S. 73° W. of Ione. Edwin-type clay in the lower member of the Ione formation. Active about 1945. Open pit. Idle. (Bates 45:5,24)
108	Smokey No. 2	Charles S. Howard Estate, c/o Crocker First National Bank of San Francisco, Post and Montgomery Sts., San Francisco. Leased to Gladding, McBean and Co., 2901 Los Feliz Blvd., Los Angeles 26, from 1948 to date.	NW¼ 34	6N (proj.)	9E	MD	Lots 283 and 284, Rancho Arroyo Seco. 2.8 miles S. 63° W. of Ione. Edwin-type clay in the lower member of the Ione formation. Active about 1945. Open pit. Idle. (Bates 45:5,24)
109	Smokey No. 3	Charles S. Howard Es-	SW¼ 34	6N	9E	MD (proj.)	Lot 279, Rancho Arroyo Seco. 3.0 miles S. 57° W. of Ione.

CLAY, (cont.)

MAP NO.	CLAIM, MINE, OR GROUP	OWNER NAME, ADDRESS	SEC.	T.	R.	B & M	REMARKS
109	Smokey No. 3 (cont'd)	tate, c/o Crocker First National Bank of San Francisco, Post and Montgomery Sts., San Francisco. Leased to Gladding, McBean and Co., 2901 Los Feliz Blvd., Los Angeles 26, from 1948 to date.	34	6N (proj.)	9E (proj.)	MD	Edwin-type clay in the lower member of the Ione formation. Active about 1945. Open pit. Idle. (Bates 45:5,24)
110	Solari (Carlile clay and sand)	E.E. Tremain, Buena Vista, via R.F.D., Ione	W½ NW¼ 8	5N	10E	MD	3.6 miles S. 25° E. of Ione. Clayey sand from the Ione Sand bed in the lower member of the Ione formation. Active from 1927 to 1949. Open pit. Idle. (Logan 23:96; 27:136-137; Dietrich 28:57-58, 262; Sampson and Tucker 31:434-35)
	Tiger Creek	Winton Lumber Company, Martell	SW¼ SW¼ 23	8N	14E	MD	Herein.
111	Wallen	Charles S. Howard Estate, c/o Crocker First National Bank of San Francisco, Post and Montgomery Sts., San Francisco. Leased to Gladding, McBean and Co., 2901 Los Feliz Blvd., Los Angeles 26, from 1948 to date.	SE¼ 6	5N	10E	MD	Lot 265, Rancho Arroyo Seco. 2.8 miles S. 23° E. of Ione. Clayey sand from the Ione Sand bed in the lower member of the Ione formation. Open pit. Idle.
112	Western Refractories Company Plant (Ione Fire Brick Company) Plant	Western Refractories. Company, Russ Bldg., San Francisco	NW¼ 31	6N	10E	MD	1.4 miles S. 18° E. of Ione. Manufacture fire brick and ground fire clay. Active. (Tucker 14:11; Boalich 20:38; Logan 23:96; 27:138-141; Dietrich 28:60-61; Sampson and Tucker 31:435; herein)

CLAY, (cont.)

MAP NO.	CLAIM, MINE, OR GROUP	OWNER NAME, ADDRESS	SEC.	T.	R.	B & M	REMARKS
113	Woolford	W.G. Woolford, R.F.D. Ione	NE¼ 11	5N (proj.)	9E	MD	2.1 miles N. 70° W. of Buena Vista. Clay of the Cheney Hill type (with a P.C.E. of Cone 33+) in the lower member of the Ione formation. Open pit and adits. Idle. (Bates 45:5-6, 24; Pask 52:16, 23)
114	Yager	Myrtle E. Yager	NW¼ 30	6N	10E	MD	0.6 of a mile S. 62° E. of Ione. Fire clay in the lower member of the Ione formation. Operated by Western Refractories Co. from approximately 1944 to 1950. Open pit. (Idle ?)
115	Yaru No. 1	Charles S. Howard Estate, c/o Crocker First National Bank of San Francisco, Post and Montgomery Sts., San Francisco. Leased to Gladding, McBean and Co., 2901 Los Feliz Blvd., Los Angeles 26, from 1948 to date.	SW¼ 4	6N (proj.)	9E	MD	Lot 232, Rancho Arroyo Seco. 4.9 miles N. 45° W. of Ione. 100 yards north of Lignite Siding on the Southern Pacific railroad. Fire clay (Yaru No. 1) and red-burning clay (Yaru No. 2) in the lower member of the Ione formation. During the 1920's the average annual production was about 4000 tons. Open pit. Idle. (Logan 27:135; Dietrich 28:56, 302,335)
116	Yaru No. 2	Charles S. Howard Estate, c/o Crocker First National Bank of San Francisco, Post and Montgomery Sts., San Francisco. Leased to Gladding, McBean and Co., 2901 Los Feliz Blvd., Los Angeles 26, from 1948 to date.	N½ 10	6N (proj.)	9E	MD	Lot 224 and 225, Rancho Arroyo Seco. 3.9 miles N. 36° W. of Ione. Open pit. Idle. (Dietrich 28:plate 6)
117	Yosemite (Harvey)	Western Refractories Co., Russ Building, San Francisco	W½ 29	7N	9E	MD	(Logan 27:138; Dietrich 28:63,298; herein)

REFRACTORY MATERIALS
GLASS SAND

MAP NO.	CLAIM, MINE, OR GROUP	OWNER NAME, ADDRESS	SEC.	LOCATION T.	R.	B & M	REMARKS
118	Custer (Grog)	E.W. Custer. Leased to Western Refractories Co., Russ Bldg., San Francisco	SE¼ 30	6N	10E	MD	Herein.
	Grog						See Custer.

o

printed in CALIFORNIA STATE PRINTING OFFICE

printed in CALIFORNIA STATE PRINTING OFFICE

84395   8-53   2M

www.ingramcontent.com/pod-product-compliance
Lightning Source LLC
Chambersburg PA
CBHW062019200326
41519CB00017B/4850